A Practical Approach to Metaheuristics using LabVIEW and MATLAB®

T0199861

A Practical Approach to Metaheuristics using LabVIEW and MATLAB®

Pedro Ponce Cruz
Arturo Molina Gutiérrez
Ricardo A. Ramírez-Mendoza
Efraín Méndez Flores
Alexandro Antonio Ortiz Espinoza
David Christopher Balderas Silva

CRC Press
Taylor & Francis Group
Boca Raton London New York

CRC Press is an imprint of the
Taylor & Francis Group, an **informa** business

A CHAPMAN & HALL BOOK

First edition published 2020
by CRC Press
6000 Broken Sound Parkway NW, Suite 300, Boca Raton, FL 33487-2742

and by CRC Press
2 Park Square, Milton Park, Abingdon, Oxon, OX14 4RN

© 2020 Taylor & Francis Group, LLC
CRC Press is an imprint of Taylor & Francis Group, LLC

ISBN: 978-0-367-49426-1 (hbk)
ISBN: 978-0-367-33704-9 (pbk)
ISBN: 978-0-429-32441-3 (ebk)

Typeset in Minion
by codeMantra

When these MATLAB and LabVIEW® programs are downloaded and used for publishing any kind of document, they have to be referenced using the book information. Also, if they are changed and adapted the corresponding credits have to be mentioned.

Contents

LIST OF FIGURES   xi

LIST OF TABLES   xvii

FOREWORD   xix

PREFACE   xxi

AUTHORS   xxiii

SECTION I   Basis

CHAPTER   1 ▪ Fundamental Concepts of Optimization   3

1.1   INTRODUCTION   3

1.1.1   Case Study of Electric Vehicle Driving and Temperature of Power Electronic Stage Optimization   6

1.1.2   Conventional Optimization   9

1.1.2.1   MATLAB Symbolic Code   12

CHAPTER   2 ▪ Software Fundamentals for Optimization   17

2.1   MATLAB FUNDAMENTALS   17

2.1.1   User Interface   17

2.1.2   Variables Definition   19

2.1.3   Constants   21

2.1.4   Arrays, Vectors, and Matrices   21

2.1.5   Basic Commands clc, who, clear, save, and load   25

2.1.6   Basic Functions   26

2.1.7   Programming   28

2.1.8   Conditionals   30

     2.1.8.1   IF   31

     2.1.8.2   Switch   31

2.1.9   Loops   32

     2.1.9.1   For Loop   32

     2.1.9.2   While Loop   34

     2.1.9.3   Break and Continue Loops   34

2.1.10   Graphs   36

2.1.11   Examples   40

2.2   SIMULINK FUNDAMENTALS   43

2.2.1   Working with Blocks   44

2.2.2   Block Settings   45

2.2.3   Simulink Models and MATLAB Variables   45

2.2.4   Simple Simulink Example   46

2.3   GENERAL INTRODUCTION TO LabVIEW   47

Section II   Metaheuristic Optimization

Chapter   3 ▪ Basic Metaheuristic Optimization Algorithms   61

3.1   EXHAUSTIVE SEARCH   61

3.2   RANDOM OPTIMIZATION   61

3.3   NELDER–MEAD ALGORITHM   61

Chapter   4 ▪ Evolution Algorithms   65

4.1   GENETIC ALGORITHMS   65

4.2   SIMULATED ANNEALING   66

4.3   TABU SEARCH   67

CHAPTER   5 ▪ Memetic Algorithms   71

5.1   ANT COLONY OPTIMIZATION   71
5.2   PARTICLE SWARM OPTIMIZATION   72
5.3   BAT OPTIMIZATION   73
5.4   GRAY WOLF OPTIMIZATION   74

CHAPTER   6 ▪ Geological Optimization   77

6.1   EARTHQUAKE ALGORITHM   78
       6.1.1   Background   78
       6.1.2   P- and S-Wave Velocities   80
       6.1.3   Earthquake Optimization Algorithm   82

CHAPTER   7 ▪ Optimization MATLAB App and LabVIEW
               Toolkit   87

7.1   MATLAB APP   87
       7.1.1   Primary User Interface   88
                7.1.1.1   Algorithm User Interfaces   89
       7.1.2   Secondary User Interface   91
                7.1.2.1   Algorithm User Interfaces   91
       7.1.3   Individual Functions   92
       7.1.4   MATLAB Simulink   96
                7.1.4.1   MPPT Simulink Models   96
7.2   LABVIEW APP - FRONT PANELS   96
       7.2.1   GA Application in LabVIEW   99
       7.2.2   PSO Algorithm Application in LabVIEW   99
       7.2.3   BA Application in LabVIEW   100
       7.2.4   ACO Algorithm Application in LabVIEW   100
       7.2.5   GWO Algorithm Application in LabVIEW   101
       7.2.6   EA Application in LabVIEW   101
       7.2.7   NM Algorithm Application in LabVIEW   102

CHAPTER 8 ▪ Equations and Ongoing Projects 103

8.1 EQUATIONS 103
8.1.1 Equation 01 103
8.1.2 Equation 02 103
8.1.3 Equation 03 104
8.1.4 Equation 04 105
8.1.5 Equation 05 106
8.2 PROJECTS 107
8.2.1 Project 01: Linear Square Regression 107
8.2.2 Project 02: Welded Cantilever Minimization 108
8.2.3 Project 03: Traveling Salesman Problem 109
8.2.4 Project 04: 3D Traveling Salesman Problem 109
8.3 MPPT CASE STUDY 111
8.3.1 Simulink Models 116
8.3.2 Results 119
8.4 INDUSTRY 4.0 CASE STUDY: THREE-PHASE INVERTER 121
8.4.1 MATLAB Optimization Solution 124
8.4.2 LabVIEW Optimization Solution 125
8.4.3 Final PCB 127
8.5 DC MOTOR SPEED CONTROLLER WITH PID TUNING OPTIMIZATION ALGORITHM 127
8.6 OPTIMIZATION ALGORITHMS EMBEDDED IN LABVIEW FPGA 132
8.6.1 Benchmark Functions 134
8.6.1.1 Implementation into FPGA 134
8.6.1.2 Benchmark Functions Utilization Summary 138
8.6.2 Optimization Algorithms Implementation 140
8.6.2.1 Pseudo Random Number Generation 141
8.6.3 Optimization Algorithms Utilization Summary 142

Appendix   143

Bibliography   147

Index   155

List of Figures

I.1   Reading book flowchart.   xxii

1.1   Metaheuristic optimization methods.   5

1.2   Topology for controlling a BLDCM by fuzzy logic and
       PID controller.   7

1.3   Closed-loop for a driving cycle mode using fuzzy logic
       controller optimized by PSO.   9

1.4   Graph of $-f_i(x)$ and $f_i(x)$.   10

1.5 $f(x)$ and $K * f(x)$.   11

1.6 $f(x)$ and $f(x) + K$.   11

1.7   Function $f(x) = (1 - x.^2) + 15/200 * x$.   13

1.8   General description of an objective function.   14

1.9   Gradient vector directions.   15

1.10   Gradient descent method.   16

2.1   MATLAB® user interface.   18

2.2   Workspace with variable ans.   19

2.3   Workspace with variable $ans = 6$.   20

2.4   MATLAB® editor.   28

2.5   Break and continue executions.   35

2.6   Graph of a cosine using plot.   36

2.7   Graph of a cosine using plot with color and line properties. 37

2.8   Graph of a cosine and sine using plot with color, line
       properties, title, labels, and legends.   38

2.9   Three-dimensional graph using the **surf()** and **meshgrid()**
       functions.   39

2.10   A Mass-Spring-Damper system modeled within Simulink®.   43

2.11   Simulink® icon.   43

2.12   (a and b) Opening Simulink® and its toolbar.   44

2.13   Integrator block properties.   46

2.14   Mass-Spring-Damper system.   47

2.15   Mass-Spring-Damper system result after it was simulated for 10 seconds in Simulink®.   47

2.16   Block diagram (a) and front panel (b).   48

2.17   Starting window.   48

2.18   The front panel menu (a) the block diagram menu (b).   49

2.19   Adding two variables by a VI program in LabVIEW, step one (a) and step two (b).   49

2.20   LabVIEW representation.   50

2.21   Path for finding functional blocks (a) and the while functional block (b).   50

2.22   The functional block "for" that uses the iteration number as stopping condition or uses the number of elements in the array to index the "for" as stopping condition.   51

2.23   Case functional block using a boolean condition.   51

2.24   Basic program using for, while, and case structures.   52

2.25   Parallel execution (a) and serial execution (b).   52

2.26   Path for getting the chart and graph plots.   53

2.27   (a and b) Chart plot using LabVIEW.   53

2.28   (a and b) Graph plot using LabVIEW.   54

2.29   Array of integers.   54

2.30   Cluster with several elements of different type.   54

2.31   Array of clusters.   55

2.32   (a and b) For loops to create a matrix.   55

2.33   (a and b) Adding operation for a matrix.   56

2.34   MathScript documentation and window into the toolbar of the front panel.   56

2.35   (a and b) Simulation of a second-order linear system using the simulation toolkit.   57

3.1   Random optimization pseudocode.   62

3.2   The four steps of the Nelder–Mead algorithm.   63

3.3   Nelder–Mead algorithm pseudocode.   64

4.1   GA next generation.   66

4.2   Genetic algorithm pseudocode.   66

4.3   Simulated annealing pseudocode.   67

4.4   Tabu search example.   68

4.5   Tabu search algorithm pseudocode.   69

5.1   Ant colony algorithm pseudocode.   72

5.2   PSO pseudocode.   73

5.3   Bat algorithm pseudocode.   75

5.4   GWO algorithm pseudocode.   76

6.1   Classification of metaheuristic optimization algorithms.   78

6.2   Origin of an earthquake.   80

6.3   Ground acceleration (three-axis) recording during an
       earthquake.   81

6.4   *P*-wave traveling through an earth material by compres-
       sion and dilation.   81

6.5   *S*-wave traveling through an earth material by shearing
       deformation.   82

6.6   Procedure to find particles within and beyond the defined
       Sr.   84

6.7   General architecture for *EA*.   85

6.8   (a and b) Example of the EA behavior with 10 epicenters.   86

7.1   Opening the main.m file running the program.   87

7.2   (a and b) Main and secondary APP.   88

7.3   Optimization algorithms: (a) NM, (b) GA, (c) simulated
       annealing, (d) PSO, (e) BA, (f) GWO, and (g) EA.   90

7.4   Example on how to use one of the applications
       (earthquake).   91

7.5   (a) Tabu search and (b) ant colony optimization.   92

7.6   TSP example with 18 cities. The algorithm (Tabu search) was run over 100 iterations.   93

7.7   Simulink® model configuration for MPPT algorithms simulation.   97

7.8   Comparison between MPPT algorithms.   98

7.9   LabVIEW APP (front panel).   98

7.10 GA algorithm with (a) sphere and (b) Styblinski–Tang (front panel).   99

7.11 PSO algorithm with (a) sphere and (b) surface functions (front panel).   99

7.12 BA with (a) sphere and (b) surface functions (front panel).   100

7.13 ACO algorithm with (a) Styblinski–Tang and (b) Keane functions (front panel).   100

7.14 ACO algorithm with (a) Styblinski–Tang and (b) surface functions (front panel).   101

7.15 EA with (a) Styblinski–Tang and (b) surface functions (front panel).   101

7.16 NM algorithm with (a) Styblinski–Tang and (b) Keane functions (front panel).   102

8.1   Project 01.   104

8.2   Project 02.   104

8.3   Project 03.   105

8.4   Project 04.   105

8.5   (a–c) Project 05 constraint boundary and contour lines in the profit maximization problem.   107

8.6   (a and b) Project 06 linear regression.   108

8.7   Project 08.   108

8.8   TSP with 18 Points.   109

8.9   3D TSP application using a pre-configured model.   110

8.10 (a and b) 3D TSP application using configured coordinates by the user.   110

8.11 Connection scheme of a power converter linked to the PV generation system. 111

8.12 Selected topology for the case study. 112

8.13 Flowchart of the P&O MPPT algorithm. 113

8.14 Flowchart of the PSO MPPT algorithm. 115

8.15 Flowchart of the EA MPPT algorithm. 116

8.16 Flowchart of the EA MPPT algorithm. 117

8.17 MPPT through time. 120

8.18 Industry 4.0. 121

8.19 (a and b) Simulated PCB. 123

8.20 Evolution soldering. This image shows the evolution of how the distance was reduced for each iteration using ACO to reduce the drilling distances and tool changes. 125

8.21 Final path for the drilling process. Each line represents the path taken for each of the different hole diameters, and the dark gray line represents a change of tool between the paths. It can be seen that for the most of the drilling processes, it finishes to drill one diameter and moves to the next diameter to find the shortest distance. This changes for the light gray where it separates the drilling into two sections and changes tool before continuing with that diameter. 126

8.22 Drill route optimization application. 126

8.23 Final PCB views: a) top and b) bottom view without components; c) PCB with motor and cable connections. 127

8.24 Block diagram of control system with PID tuning optimization. 128

8.25 General structure of control system with PID tuning optimization. 129

8.26 DC motor speed controller (front panel). 130

8.27 DC motor speed controller (manual mode for plant identification). 131

8.28 Simulation of control system in closed loop. 131

8.29 DC motor speed controller (automatic mode with PID gains obtained by metaheuristic algorithms). 132

8.30 DC motor speed controller (remote application and monitoring by video streaming). 132

8.31 National Instruments cRIO-9030 FPGA specifications. 133

8.32 Surfaces of the benchmark functions used to compare the performance of algorithms [34,48]. 135

8.33 Memory used as a 2D-LUT. 138

8.34 Sub-VI implemented to calculate $sin(x)$ function. 139

8.35 Sub-VI implemented to calculate $cos(x)$ function. 140

8.36 VI of Keane function. 140

8.37 Sub-VI of pseudorandom number generation. 141

8.38 Device utilization summary (%). 142

A.1 Graph of 13 cities arranged in a rectangle. The salesperson starts at the origin and visits all 13 cities once and returns to the starting point. The obvious solution is to trace the rectangle, which has a distance of 14. 144

A.2 Schematic design of the implemented H-bridge. 144

A.3 Designed testbed. 145

A.4 Schematic of the implemented test circuit. 145

A.5 Board design of the implemented test circuit. 146

A.6 Testing environment. 146

List of Tables

2.1   Mathematical Operations in MATLAB®   23

2.2   Elementary Mathematical Functions in MATLAB®   26

2.3   Common Styles used in MATLAB®   38

6.1   Principal Materials or Geologic Formation Properties [7]   82

8.1   Regression Points   107

8.2   Converter Parameters   118

8.3   PV Array for Low-Power Simulation Parameters   119

8.4   Simulations Parameters Used in MATLAB® Simulink®   119

8.5   Design Parameters of the Inverter   122

8.6   Selected Components for the Design   124

8.7   Required Drills   124

8.8   Benchmark Functions Implemented in the FPGA Used to Evaluate Each Optimization Algorithm   136

8.9   FPGA Utilization Summary with Benchmark Functions   141

8.10 FPGA Utilization Summary with Optimization Algorithms (Sphere as Objective Function)   142

Foreword

Optimization is the procedure used to determine the variables of a function that will maximize or minimize the function value. The variables of the functions are often governed by constraints that limit the search space to determine the optimal values for the function. Optimization problems are found in the real world wherever there is a limit on resources, such as time, money, raw materials, and energy, that are necessary to maximize or minimize certain outcomes. The study of optimization techniques is crucial to solving a wide range of problems in multiple areas, including engineering, economics, artificial intelligence, sociology, geology, and genetics.

Real-world problems in optimization are highly complex with nonlinearities, large numbers of variables, and non-convexity. Optimization techniques are myriad and diverse. They can be classified into gradient-based approaches (e.g., hill climbing) and non-gradient-based approaches (e.g., Nelder–Mead method). They can also be classified into deterministic techniques (e.g., hill climbing) and stochastic techniques (e.g., genetic or particle swarm optimization). In terms of search strategy, the algorithms can be categorized as local and global. Since local search algorithms are straightforward and efficient, they can yield locally suboptimal solutions. Global search algorithms can explore the solution space and yield global solutions. Such techniques are termed as heuristic. Metaheuristic approaches are an extension of heuristic approaches where a strategy is applied to guide the heuristic to exploit randomization and local search. Metaheuristic approaches can be further classified into single-solution approaches (e.g., simulated annealing) and population-based (e.g., genetic) approaches. Further, population-based approaches are organized into multiple categories. This book presents summaries of the most important metaheuristic approaches along with examples whose solutions are illustrated using their toolbox.

This book outlines a toolbox to solve optimization problems using a host of metaheuristic approaches, including genetic algorithms, ant colony optimization, particle swarm optimization, Nelder–Mead optimization, and earthquake optimization. The toolbox has a LabVIEW front end and a MATLAB® engine. This book provides a brief introduction to the different metaheuristic algorithms in the toolbox along with solutions for various examples followed with illustrations. The authors do not presume the readers to be experts at MATLAB and LabVIEW and dedicate significant parts of their book to helping the reader using MATLAB, Simulink®, and LabVIEW. As the title states, this book is a practical guide to applying metaheuristics to solve optimization problems. It will be a useful resource for students wanting to gain hands-on experience with metaheuristic optimization and a handy reference for experts as well.

Hemanth Kumar Demakethepalli Venkateswara
**School of Computing, Informatics, and Decision Systems Engineering
Asst Research Professor (FSC)**
Arizona State University
Faculty, TEMPE Campus, Mailcode 8809

Preface

HOW TO READ THIS BOOK

The flowchart in Figure I.1 is a guide to how to read this book in a fast-track method: Just select the topic you want to learn more about and follow the diagram.

All the LabVIEW and MATLAB® programs presented in the chapters of this book can be downloaded at www.crcpress.com/9780367337049.

The password of the zip programs is $BOOK-MATLABOpt2020\#$.

MATLAB® is a registered trademark of The MathWorks, Inc. For product information, please contact:

The MathWorks, Inc.
3 Apple Hill Drive
Natick, MA 01760-2098 USA
Tel: 508-647-7000
Fax: 508-647-7001
E-mail: info@mathworks.com
Web: www.mathworks.com

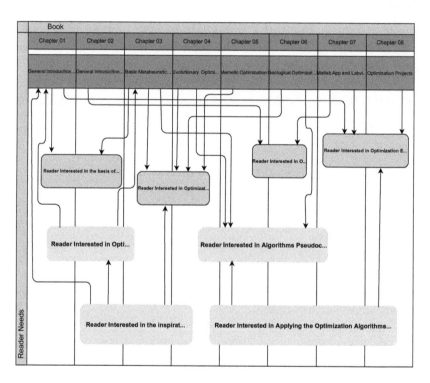

Figure I.1 Reading book flowchart.

Authors

 **Dr. Pedro Ponce Cruz** is a senior researcher scientist and member of the Product Innovation Research Group at Tecnológico de Monterrey, Mexico. He received his B.Eng. degree in automation and control, and his M.Sc. and Ph.D. degrees in electrical engineering focused on automatic control systems from Instituto Politecnico Nacional, Mexico, Mexico City. He worked for more than 12 years as a field and design engineer in control systems and automation. He has also authored and co-authored more than 100 research articles, 14 books, and 6 inventions, including one US Patent. His research interests include mechatronics, control systems, smart grids, machine learning, product design, optimization, electrical engineering, education, and artificial intelligence. He is a member of the Mexican National System of Researchers and Mexican Academy of Science. Besides, he is an associate editor of the *International Journal of Advanced Robotics Systems* and the *International Journal on Interactive Design and Manufacturing.*

 **Dr. Arturo Molina Gutiérrez** is the vice president of Research and Technology Transfer and a professor at Tecnológico de Monterrey University, Mexico. He received his B.Eng. degree in computational systems and his M.Eng. degree in computational sciences from Tecnológico de Monterrey University, his Ph.D. in manufacturing engineering from Loughborough University of Technology, and his Ph.D. in mechanical engineering from the Technical University of Budapest, Hungary. He has co-authored more than 300 research articles and several inventions. His current research interests include concurrent engineering

technologies and manufacturing, enterprise integration engineering, and technology management. He has applied research and development for the economic sectors of automotive, construction, and energy. He is a member of the Mexican National System of Researchers, the Mexican Academy of Science, and the Mexican Academy of Engineering.

Dr. Ricardo A. Ramírez-Mendoza received his Ph.D. in automation from the Institut National Polytechnique de Grenoble (INPG), France, in 1997. He is now a professor of mechatronics and mechanics engineering and the dean of research, National School of Engineering and Science, Tecnológico de Monterrey within the School of Engineering and Science, Mexico. His main research interests include the applications of advanced control to automotive systems, fault detection and isolation (FDI), electromobility, biomedical signal processing, and engineering education. He is the (co-)author of 3 books, more than 100 papers in top journals, more than 170 international conference papers, and more than 1,000 citations. He has also worked as an expert consulting for different industries and regional development projects.

Efraín Méndez Flores received his B.S. degree in mechatronics engineering from Tecnológico de Monterrey, Mexico.

After receiving his B.S. degree, he was in charge of electronics software and hardware design between 2016 and 2019 as part of a startup team dedicated to the design of power inverters for the control of three-phase induction motors.

As part of Project 266632 "Laboratorio Binacional para la Gestión Inteligente de la Sustentabilidad Energética y la Formación Tecnológica" ("Bi-National Laboratory on Smart Sustainable Energy Management and Technology Training"), funded by the CONACYT (Consejo Nacional de Ciencia y Tecnología) SENER (Secretaría de Energía) Fund for Energy Sustainability (Agreement S0019201401), he joined the Ph.D. in engineering sciences program at Tecnológico de Monterrey in 2016.

He is currently involved in PV (photovoltaic) and power electronics research topics at Tecnológico de Monterrey, specifically dealing with the development of MPPT (maximum power point tracking) through metaheuristic algorithms.

Alexandro Antonio Ortiz Espinoza is a Ph.D. candidate in power electronics, energy harvesting, and renewable energy from Tecnológico de Monterrey, Mexico. He received his B.S. and MSc degrees in automatic control.

He has currently 3 years of teaching experience and 5 years of industrial experience focused on FPGAs.

Dr. David Christopher Balderas Silva was born in Mexico City, Mexico. He received his B.Eng. degree in mechatronics engineering from Universidad Panamericana in 2005, his M.S. degree in biomedical engineering from Delft University of Technology in 2009, and his Ph.D. degree in engineering sciences, with specialization on artificial intelligence and brain–computer interfaces, from Tecnológico de Monterrey CCM (campus Mexico City) in 2018.

Since 2014, he has been an assistant professor and a researcher at Tecnológico de Monterrey, Mexico, teaching more than ten different subjects. He is the author of two book chapters and eight research articles. His research interests include artificial intelligence – especially neural networks, classification, optimization, parameter estimation, and reinforcement learning – and biomedical applications such as brain–computer interfaces, haptics, and prosthesis.

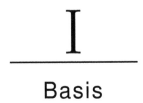

Basis

CHAPTER 1

Fundamental Concepts of Optimization

1.1 INTRODUCTION

When the simplex method for solving linear programming problems was created in the 1940s, it opened an innovative alternative for solving optimization problems in several areas such as engineering, medicine, economics, and science, among others. Later in 1984, Newton-based optimization methods were used for solving complex electric problems [49]. In addition, the development of computers allows creating novel optimization methodologies. Currently, optimization algorithms can be used for solving several complex optimization problems; this complexity is determined based on the actual and practical nature of the objective function or the model constraints [3]. Optimization algorithms are applied in several fields at the beginning of the solving problems in traditional methods using derivative-based techniques. However, those techniques sometimes are trapped in local minima and limited to appropriate specific optimization functions. There are two fundamental conditions when an optimization search is conducted: exploration and exploitation. Exploration is when it is looking for the solution into the complete new regions of the search space, and exploitation is defined as searching into regions of a search space within the previously visited points. As a result, an optimization algorithm is used to find a tradeoff between exploration and exploitation [11]. Consequently, heuristic optimization methodologies have been developed in order to deal with problems that are not solved by derivative methods [3]. As a result, a great number of heuristic and metaheuristic algorithms have been proposed [12]. Heuristic optimization methodology is trying to achieve a good and feasible solution using a trial-and-error evaluation; usually, this solution has to be found in a short period of time. It is very useful when all the solutions or combinations on a search space cannot be evaluated. Thus, the best solution is frequently not found. The heuristic algorithms are expected to work most of the time, but not all the time. On the other hand, metaheuristic algorithms use tradeoff randomization and perform a local search for trying to find a global best solution. Heuristic algorithms are popular when it is searched for a sub-optimal solution. Alan Turing developed a heuristic algorithm that is used for breaking German Enigma ciphers during World War II, which is considered as the first application. This algorithm can find the correct combination in a reasonably practical time. Turing named this algorithm a heuristic search [86]. Simulated annealing, which is one of the most representative metaheuristic algorithms, was proposed by Scott Kirkpatrick, C. Daniel Gelatt, and Mario P. Vecchi in 1983; this algorithm is applied in annealing process of metals. This method allows us to get out of a local optima and includes a probability factor [81]. Later in 1986, Fred Glover presented a methodology called Tabu search; in 1992, ant colony optimization algorithm was developed; although these algorithms are widely known, in 1995, J. Kennedy and Russell created one of the most important methodologies of metaheuristic optimization called particle swarm optimization (PSO). PSO is an optimization algorithm inspired by swarm intelligence of birds and fish, which allows solving complex optimization problems. It is evident that PSO is better than traditional search algorithms and even better than genetic algorithms [86]. Metaheuristic algorithms for solving optimization problems is generally classified into four main divisions, which were presented in [65]; this classification has been modified since a new novel methodology called earthquake optimization was proposed in 2017, which was inspired by the nature of earthquakes. Observing nature, it is an excellent alternative for creating new optimization methods; in this case, an algorithm was proposed based on the physical phenomenon, namely, the earthquake that devastated the entire Mexico City [60]. This initial algorithm was implemented and modified in [45]. As seen, there are several metaheuristic algorithms that have been successfully implemented. However, there are some challenges that have to be tackled such as accuracy when they deal with incomplete datasets, which have not been completely solved yet. For example, classification accuracy is linked with the learning algorithm method. Hence, the same dataset applied to different algorithms achieved dissimilar classification accuracy labels; other challenges are stability, scalability, and computational cost [42]. To reduce the computational cost, surrogate evaluation has been proposed and a Markov network model has been implemented as a surrogate fitness function for a genetic algorithm to form a new algorithm called Markov fitness model genetic algorithm [42]. Thus, a decrement in the computational cost is achieved [42]. A general classification of inspired metaheuristic algorithms can be done using four sets: evolution-based methods, physics-based methods, swarm-based methods, and human-based methods [65]. When laws of nature evolution are used for inspiring, evolution-based metaheuristic algorithms can be defined [65]; on the other hand, there are new optimization methods that are implemented for specific problems such as seeker optimization algorithm (SOA), social-based algorithm (SBA), and group counseling optimization (GCO) algorithm [45]. A general description of the optimization methods is presented in Figure 1.1.

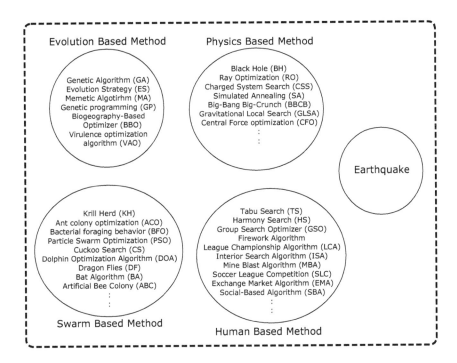

Figure 1.1 Metaheuristic optimization methods.

On the other hand, when D. H. Wolpert and Macready published the no free lunch theorems for optimization, it was well defined that there is not a universal better algorithm for all the problems so the idea of creating a better algorithm for all the problems is not a valid one; it is better to create an algorithm for most types of problems. For instance, if the performance of algorithm A is better than that of algorithm B for an optimization function, then B will outperform A for other functions. While considering properties of the algorithms, robustness, efficiency, and accuracy will be taken into account. Robustness allows us to find good solutions for a variety of problems from the same class. Efficiency restricts the computational resources such as computational time and/or storage. Finally, accuracy tries to avoid being extremely sensitive to errors in the data or calculation errors caused by rounding when a digital calculation is used. Nevertheless, generally, these characteristics are conflicted each other [58].

In fact, not only do the metaheuristic algorithms allow us to combine with several techniques like fuzzy logic systems [64] to solve control problems that require an optimization algorithm but also they can be implemented online using specialized hardware like FPGAs (field-programmable gate arrays). For giving an idea about the application of metaheuristic optimization method to solve problems, an example using electric cars is described below, in which a metaheuristic optimization method was proposed for extended lifetime in the power electronic stage of an electric vehicle without decrement in the driving performance.

1.1.1 Case Study of Electric Vehicle Driving and Temperature of Power Electronic Stage Optimization

The number of electric vehicles is increasing exponentially around the world, so electric motors play an important role and several companies are creating new materials for designing those vehicles. The first generation of electric vehicles designed with DC motors are not good enough for electromobility applications due to the lack of high efficiency and larger weight. Hence, DC motors were replaced by AC motors and induction motors. Nowadays, brushless DC motors (BLDCM) are implemented since they have important advantages in electric vehicle applications, such as lightest weight and high efficiency providing improved driving range [82].

On the other hand, the power electronic topology for controlling BLDCM is based on a three-leg inverter [61] that is usually designed with semiconductors such as an insulated-gate bipolar transistor (IGBT) and field-effect transistor (FET). A combination of PID (proportional integral derivative) and fuzzy logic speed controllers has also been deployed [57] for closing the speed loop. As shown in Figure 1.2, a fuzzy logic speed controller and a PID controller are integrated into the mechanical braking force in an electric vehicle. This kind of application that comprises conventional and intelligent controllers is typically presented. Nevertheless, this application does not tackle the decrement in the lifetime of the power electronic stage. Additionally, the sensing system uses Hall effect sensors, which acquire the rotor position and provide an electrical signal according to the magnetic field strength. The control systems such as fuzzy logic, fuzzy-PSO, and PID are commonly used in the speed control design of BLDCM electric motors. The speed control design of BLDCM requires a high-power switching in the semiconductors. Thus, the semiconductors operate under high thermal stress to achieve the speed reference. As a result, the optimization targets are the temperature and speed. The first one is related to the power electronic stage, and the other one is directly related to the BLDCM. Moreover, the two variables are linked to the speed control system.

Sometimes, after a certain period of operation, the semiconductors are damaged due to the speed controller that stresses the power electronic devices, when the controller generates command signals that are unregulated, which provokes that their lifetime is reduced [22]. Hence, a great effort for predicting and improving the lifetime in semiconductors has been being pushed with the purpose of developing new control algorithms that will help to improve the conditions within the semiconductor. Normally, the study of the lifetime of power electronic devices is carried out through real experiments by power electronic manufacturers that provide information to design power electronic stages with

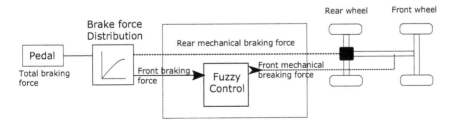

Figure 1.2 Topology for controlling a BLDCM by fuzzy logic and PID controller.

components such as IGBTs or FETs. During these experiments, the evaluation conditions are repetitively applied until the device blows up. As a result, the capacity of power cycling and temperature cycling is provided by power electronic component manufacturers. These experiments help to predict failures; however, during real operation, the power electronic stage works under different conditions such as temperature, voltage, and current, and depends on the command signals from the controller. Due to this fact, the control design stage plays an important role in reducing early failure and increases the reliability of the power electronic stage, which works under several operating conditions such as thermal stress, current, and voltage. Integration of a speed controller into BLDCM achieves the driving requirements as well as the minimization of the thermal conditions in the power electronic stage, which maximizes the lifetime. Hence, the designed controller uses an optimization process based on an objective function that considers the temperature of the power electronic stages to increment their lifetime and track motor speed reference. In fact, the motor speed response and the temperature in semiconductors are considered in the proposed objective function for tuning the fuzzy logic controller in order to increase the lifetime of power electronic devices. Co-simulations are used for validating the optimization of power electronic lifetime and track of speed BLDCM reference in dynamic regions; co-simulation is carried out by software and hardware elements. Co-simulation helps to combine two or more specialized programs utilizing advanced models and avoiding implementation problems between the control systems and power electronic stage [60]. Co-simulation consists of coupled and unified two-system modeling in different languages in order to analyze them as a unique system, which allows simulating dynamic scenarios, and to analyze their behavior during certain periods of time [22]. When a new electric drive system is designed, it is important to know its global and local temperature behavior as well as its operation conditions. On the other hand, the PSO algorithm has been implemented in several fields [62], because it has a few parameters that need to be tuned and it is easy for implementation. For solving this optimization problem, the proposed control scheme is illustrated in Figure 1.3, in which a fuzzy controller is optimized by a metaheuristic algorithm (PSO). Besides, optimization algorithms define the objective function as the fundamental step, and the proposed algorithm defines the cost function as the sum of error of the thermal condition into the power electronic stage e_t and error of vehicle speed e_c, $f_c = e_c + e_t$, Figure 1.3. The errors are calculated as the difference

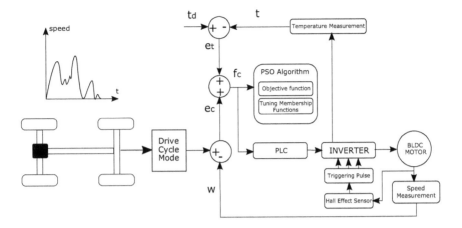

Figure 1.3 Closed-loop for a driving cycle mode using fuzzy logic controller optimized by PSO.

between the desired value and the current value. In Ref. [43], it is shown that a fuzzy logic controller optimized with PSO is used for expanding the lifetime of power electronic devices into BLDCM drives so it is possible to use the same cost function.

1.1.2 Conventional Optimization

Mathematically speaking, an optimization problem can be solved as shown in equation 1.1.

$$
\begin{aligned}
f_i(x) \quad & (i = 1, 2, \ldots, G) \\
n_j(x) \leq 0 \quad & (j = 1, 2, \ldots, l) \\
h_k(x) = 0 \quad & (k = 1, 2, \ldots, Q)
\end{aligned}
\tag{1.1}
$$

Minimize $x \in R^n$, where R^n is the search space; $f_i(x)$ is the objective function, which is also named energy function or cost function; $x = (x_1, x_2, \ldots, x_n)^T$ are the decision variables, which can be discrete, continuous, or a mixture of the two; n_j and h_k are the constraints, which are generally called inequalities or equalities.

The cost function can be linear or nonlinear, and if the constraints are linear, the problem is linearly constrained. Thus, if the cost function and the constraints are linear, it is considered as a linear programming problem. Nevertheless, if the cost function and constraints are nonlinear, it is considered as a nonlinear optimization problem. On the other hand, if there are no constraints, the only task is to find the minimum or

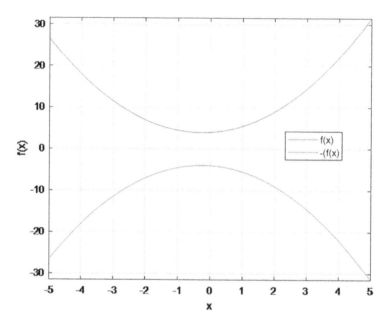

Figure 1.4 Graph of $-f_i(x)$ and $f_i(x)$.

maximum of the objective function. Figure 1.4 illustrates the general description of an objective function. Sometimes, the objective function consists of more than one maximum or minimum. The global maximum is the largest value of the function at its complete domain, whereas the global minimum is the smallest value of the function at its entire domain. Thus, the local maximum is a high point or upward peak at its domain, but not the highest one, whereas the local minimum is a low point or downward peak at its domain, but not the lowest one. Since it is difficult to reach the global minimum or maximum, in most cases it is good enough to find the local minimum or maximum of the cost function into its neighborhood. The maximization is equivalent to the minimization of $-f_i(x)$, and any inequality $Z_j(x) \leq 0$ is equivalent to $-Z_j(x) \geq 0$. Below, the symbolic code showing an example of $-f_i(x)$ and $f_i(x)$ is described. Figure 1.2 shows the graph of these two functions. There are basic mathematical operations such as multiplication $k * f(x)$ and addition $k + f(x)$ of a constant that do not affect the optimum solution point. Figures 1.4–1.6 illustrates the operations that are evaluated using MATLAB®[1].

---

[1]MATLAB coding will be explained in the following chapters.

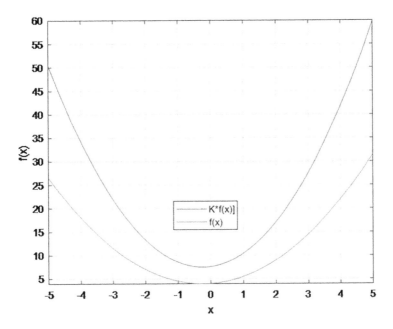

Figure 1.5 $f(x)$ and $K * f(x)$.

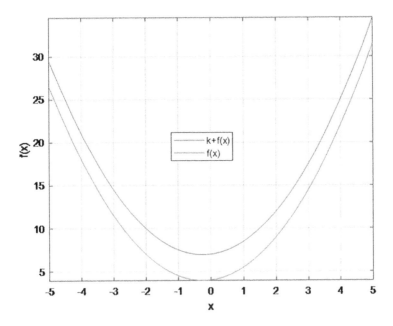

Figure 1.6 $f(x)$ and $f(x) + K$.

1.1.2.1 MATLAB Symbolic Code

```
syms t
y = t^2+(0.5*t)+4;
x = t;
fplot(x,y)
grid
hold
y = -1*(t^2+(0.5*t)+4);
x = t;
fplot(x,y)
```

MATLAB code

```
clear all
clc
syms t
k=1.9
y = k*(t^2+(0.5*t)+4);
x = t;
fplot(x,y)
grid
hold on
syms t
y = t^2+(0.5*t)+4;
x = t;
fplot(x,y)
hold off
```

MATLAB code

```
clear all
clc
syms t
k=3
y = k+(t^2+(0.5*t)+4);
x = t;
fplot(x,y)
grid
hold on
syms t
y = t^2+(0.5*t)+4;
x = t;
```

```
fplot(x,y)
hold off
```

In general terms, for finding the maximum value of a function in a complete mesh, it is possible to find it using MATLAB code presented below. Figure 1.7 depicts the function, and the maximum point is determined using MATLAB code.

MATLAB code

```
f=inline('(1-x.^2)+15/200*x')
x=-2:0.0001:2;
[z,i]=max(f(x))
x0=x(i)
xd = @(x) (1-x.^2)+15/200*x;
fplot(xd)
grid
```

Results

```
f =
#Inline function:
f(x) = (1-x.^2)+15/200*x
z =1.0014
i = 20376
x0 = 0.0375
```

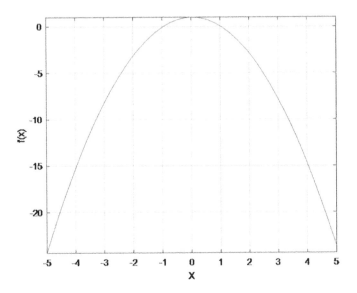

Figure 1.7 Function $f(x) = (1 - x.^2) + 15/200 * x$.

The point x^* is called a weak local minimizer, which can be expressed as follows [58]:

If there is a neighborhood N of x^*, then $f(x^*) \leq f(x)$ for all $x \in N$.

A strict local minimizer is defined as the minimum value into its neighborhood and is called the strong local minimizer [58]: If there is neighborhood N of x^*, then $f(x^*) < f(x)$ for all $x \in N$ with $x^* \neq x$.

Figure 1.8 depicts the local and global minimum of a function $f(x)$.

If the function is smooth and twice differentiable, it is not indispensable to search all the points in the immediate vicinity of x^* for finding the smaller function value and it uses only the gradient $\nabla f(x^*)$ and the Hessian $\nabla^2 f(x^*)$.

When a function is smooth, the gradient vector $\nabla f(x)$ at x is perpendicular to the surfaces of constant function value [58]. The gradient is defined by first-order partial derivatives, for a function $f(x)$. See Figure 1.9.

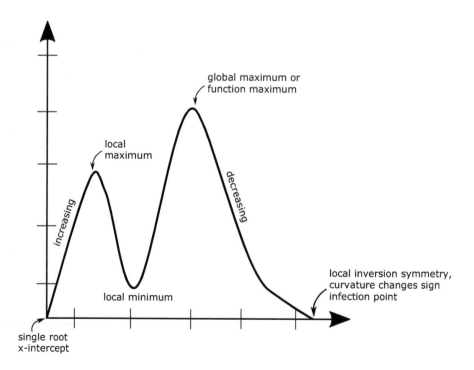

Figure 1.8 General description of an objective function.

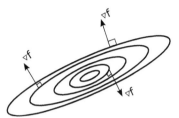

Figure 1.9 Gradient vector directions.

$$\nabla f(x) = \begin{bmatrix} \dfrac{\partial f(x)}{\partial x_1} \\ \dfrac{\partial f(x)}{\partial x_2} \\ \vdots \\ \dfrac{\partial f(x)}{\partial x_n} \end{bmatrix}$$

A critical point for a multivariable function $f(x, y)$ is a point $f(x_0 + y_0)$ in its domain where f is differentiable with $f_x(x_0 + y_0) = 0$. When the function is twice differentiable at point x, the Hessian matrix is defined by second-order partial derivatives.

$$H(x) = \begin{bmatrix} \dfrac{\partial^2 f(x)}{\partial x_1^2} & \dfrac{\partial^2 f(x)}{\partial x_1 x_2} & \vdots \\ \dfrac{\partial^2 f(x)}{\partial x_2 x_1} & \dfrac{\partial^2 f(x)}{\partial x_2^2} & \vdots \\ \vdots & \ddots & \vdots \\ \dfrac{\partial f(x)}{\partial x_n x_1} & \cdots & \dfrac{\partial f(x)}{\partial x_n^2} \end{bmatrix}$$

If several multi-objective functions have to be minimized at the same time, Pareto optimality has to be used instead of the concept of optimality.

There are different conventional methods used for solving optimization problems; some of those algorithms are presented below [49]:

- When there are linear constraints and a linear objective function with non-negative variables, the simplex method can be used. Mixed-integer programming method that belongs to a type of linear programming can be implemented if the variables of constraints are restricted to be integers.

- If there are constraint functions with equality and/or inequality and a nonlinear objective function, a nonlinear programming algorithm can be implemented.

- If the objective function is a quadratic function with linear constraints, quadratic programming can be implemented.

- When the conditions of optimality are required (Kuhn–Tucker conditions), an iterative method consisting of nonlinear equations can be implemented.

It is difficult to find analytic solutions, but it is possible to apply iterative algorithms such as descendent direction methods.

In general terms, the gradient descent method can be implemented using an iterative algorithm as illustrated in Figure 1.10. Where C is the step size in the search space and k is the current iteration,

$$x^{k+1} = x^k - C\nabla f(x)$$

In the next chapters, a complete review of the main metaheuristic optimization algorithms as well as some applications to allow readers to implement those algorithms in experimental or simulated conditions is presented. The aim of this book is to provide new information for the implementation and application of metaheuristic optimization. Since this book does not show a strict review of conventional optimization, it is recommended to review some of the references that are used in this chapter for learning more into conventional optimization.

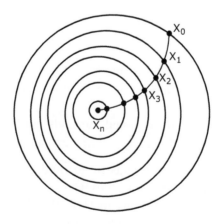

Figure 1.10 Gradient descent method.

CHAPTER 2

Software Fundamentals for Optimization

2.1   MATLAB FUNDAMENTALS

MATLAB® is a software oriented to the realization of mathematical projects that include complex calculations or are of great magnitude, and allows its visualization. MATLAB is a contraction of Matrix Laboratory, as it was initially developed for handling matrices for linear algebra.

Nowadays, MATLAB is used in a large number of universities since it is a useful tool to teach courses such as linear algebra, image processing, differential equations, complex analysis, control theory, and artificial intelligence. Currently, it is also one of the most used software packages for science since it has a high mathematical capabilities, many packages, many user-developed implementations, an interaction with several kinds of software and hardware elements, and modeling capabilities.

2.1.1   User Interface

The main way to interact with MATLAB is its user interface (see Figure 2.1). The interface has to be friendly so quick access icons for folders and other usability elements have to be integrated on it. The main elements are listed below:

- Ribbon, where the main quick access icons can be found with some tabs to separate depending on what is going to be used

- Current Folder, which consists of the current folder address located below the Ribbon and the current folder window that shows the folders and files contained inside the current folder address

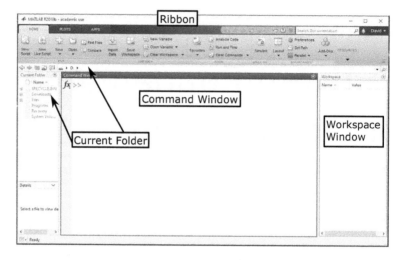

Figure 2.1 MATLAB® user interface.

- Workspace window, where the variables that have been currently defined in the workspace will be displayed.

Once MATLAB has finished its starting procedure, it is ready to start working using the MATLAB prompt that will be enabled in the command window (double arrow, $>>$).

The command window in MATLAB is used to perform many calculations in a manner similar to a scientific calculator. For example, to multiply five times 3, it just requires to type

$>> 5 * 3$

The same can be done for more complex operations. For example, to find sine of $\pi/2$, it just requires to type

$>> sin(pi/2)$

If [Enter] is pressed after typing any of two commands, the result will be shown using the prefix ans (short for answer). For instance, in the previous two examples, the output will be $ans = 15$ and $ans = 1$, respectively. This is because these results are not assigned to any variable, so MATLAB assigns them to the variable ans. This can be observed at the workspace where the variable ans can be seen with the value of the last operation (Figure 2.2).

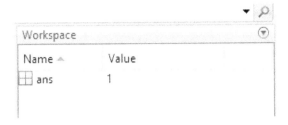

Figure 2.2 Workspace with variable *ans*.

The signs $+, -, *, /$ and $^$ denote the arithmetic operations of addition, subtraction, multiplication, division, and elevation to a power, respectively. MATLAB also uses the parentheses () to concatenate expressions, but not the brackets that are reserved for entering vectors and matrices, or brackets that are needed for a more complex type of variable not covered here. For example,

>> $3 * ((1 + 3)^(1/2))$

gives the answer, $ans = 6$. At this moment, it is important to express that MATLAB uses standard algebraic rules for its operations:

1. Brackets: Perform calculations inside brackets, working from the innermost set to the outermost set.

2. Orders: Perform power and root operations.

3. Division and multiplication: Perform multiplication and division operations, working from left to right.

4. Addition and subtraction: Perform addition and subtraction operations, working from left to right.

2.1.2 Variables Definition

MATLAB variables represent the values stored in the computer memory. These variables are available for use and can be overwritten at any point. As previously stated, if nothing is declared, MATLAB creates the variable *ans* and indicates its current value in the workspace. For example, for the operation $3 * ((1 + 3)^(1/2))$, the workspace will show the result that the variable *ans* has the value of 6, which is illustrated in Figure 2.3, where the variable has a value of 6, a size of 1×1, and the

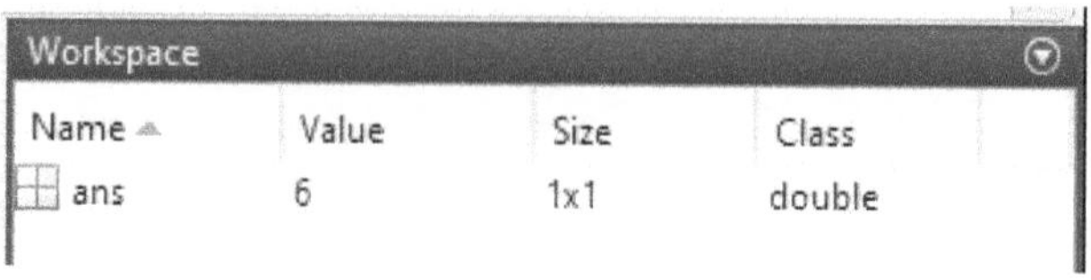

Workspace			
Name ▲	Value	Size	Class
ans	6	1x1	double

Figure 2.3 Workspace with variable $ans = 6$.

class double (to make the same properties available, right-click on Name and define which ones are visible).

As previously explained, MATLAB creates a variable ans every time any command is correctly typed and the [Enter] is pressed. Yet in MATLAB, custom variables can be created and are assigned naturally. It is just necessary to write a variable name, then the = sign followed by the value the variable will take. To accept, as always, press [Enter]. While writing only the name of a previously assigned variable, MATLAB returns its value. For example, to assign two variables:

$$>> a = 4, b = 2$$

and press [Enter]. Then, operations can be typed as

$$>> a * b$$

or to assign to a new variable

$$>> c = a * b$$

At this moment, the result of operations always appears after pressing [Enter]; if this result wants to be suppressed, a semicolon (;) can be used after the expression. Obviously, using the semicolon does not affect the variable definition, but just avoids overcrowding the command window with intermediate results. For example,

$$>> a = 4; b = 2;$$

$$>> c = a * b;$$

$$>> d = c\text{^}2 * b - a$$

will only display

$$d =$$

$$124$$

2.1.3 Constants

MATLAB has certain variables defined for special purposes, for instance, i or j, which is designated as the imaginary unit. The number π is defined using the variable pi. The e-number is not assigned, but can be obtained using the function $exp(1)$. Another important defined number is eps, which indicates the relative precision in floating point operations (highest positive that satisfy $1 + eps = 1$).

There are also other important numbers that are useful to remember in MATLAB, such as Inf and NaN that stand for the operations $1/0$ and $0/0$ and are termed as infinite and not a number. Most operations using these variables tend to produce the same variable; for example, $>> Ing * 3$ will result in Inf, and $>> NaN + Inf$ will result in NaN.

2.1.4 Arrays, Vectors, and Matrices

In MATLAB, most variables are data arrays that are translated into an order set of the same type of variable. This way, a variable can have many historical values of certain phenomena (e.g., car velocities on a trip).

A vector is a one-dimensional array that can be either a row or a column array. Row vectors are represented by brackets, and for the separation of each element, commas or spaces are used. For example,

$$>> p = [3, 4, 5 \quad 6]$$
$$p =$$

    3   4   5   6

Note that in the aforementioned example, both commas and spaces were used to define the array and no error occurred.

On the other hand, column vectors are defined by brackets, but for the separation of each element, semicolon or a carriage return is used. For example,

$$>> f = [3; 4; 5$$
$$6]$$
$$f =$$

    3
    4
    5
    6

Another way to create a vector is using double dot (:). It uses the initial value, the step size, and the final value.

$$>> c = 1.1 : 0.5 : 3.1$$
$$c =$$

    1.1000   1.6000   2.1000   2.6000   3.1000

Note that if the step size is omitted, MATLAB will use a step of one.

$$>> c = 1.1 : 3.1$$
$$c =$$

    1.1000   2.1000   3.1000

To access an element on a vector, the element has to be placed between parentheses after the variable. For example, $>> c(3)$ will result in 2.1000. Also, more than one element can be selected. For instance, selecting the last 3 elements in the vector can be done using $>> c(3 : end)$, which will result in 2.1000   2.6000   3.1000.

Analogous to a vector, two-dimensional arrays are mathematical matrices, which have many applications in science and engineering. Matrices are written between brackets, elements in a row are separated using commas and spaces, and a new row is created by a semicolon or by pressing the [Enter]. For example,

$$>> A = [1, 2, 3; 4, 5, 6; 7, 8, 9]$$
$$A =$$

    1 2 3
    4 5 6
    7 8 9

All rows must have a consistent number of elements; in other words, the number of elements per row will be the same, which is also a requirement for a higher-dimension array. Similar to vector, the matrix can access just one element instead of two elements. For example, if the matrix accesses the element belonging to the third row and second column, then it can be written as $>> A(2, 3)$, which will result in 8. Also, similar to vectors, matrices can access more than one element at a time; for instance, if the matrix accesses the second element and the third element from the second row, then it can be written as $>> A(2, 2 : 3)$, which results in 5   6.

A key point of MATLAB is that it follows linear algebra rules, and some of these rules are summarized in Table 2.1.

TABLE 2.1   Mathematical Operations in MATLAB®

Mathematical Operation	Example
	$>> A = [1, 2, 3; 456]$;
	$>> A'$
Transpose: Rotates the matrix over the main diagonal (columns become rows, and vice versa).	$ans =$
	1 4
	2 5
	3 6
	$>>A = [1, 2, 3; 4, 5, 6]$;
	$>>B = [7, 8, 9; 1, 1, 3]$;
	$>>A + B$
	$ans =$
Addition and subtraction both require matrices that have the same sizes, and these operations are performed element-wise. Note that the subtraction is done with transpose matrices.	8 10 12
	5 6 9
	$>>A' - B'$
	$ans =$
	−6 3
	−6 4
	−6 3
	$>>A = [1, 2, 3; 4, 5, 6]; s = 5$;
	$>>A * s$
Scalar multiplication of each element is multiplied by the scalar	$ans =$
	5 10 15
	20 25 45
	$>>A = [1, 2, 3; 4, 5, 6]$;
	$>>B = [7, 8, 9; 1, 1, 3]$;
	$>>A * B$
	Error using *
Matrix multiplication requires that the inner dimensions are the same. Note that for the matrix to be possible in the second case, the second matrix has to be transposed.	Incorrect dimensions for matrix multiplication...
	$>>A * B'$
	$ans =$
	50 12
	122 27

(Continued)

TABLE 2.1 (*Continued*) Mathematical Operations in MATLAB®

Mathematical Operation	Example
	$>>A = [1, 2, 3; 4, 5, 6]; s = 5;$
	$>>A/s$
Scalar division of each element is divided by the same scalar	$ans =$
	0.2000 0.4000 0.6000
	0.8000 1.0000 1.2000
	$>>A = [1, 1, 1; 0, 2, 5; 2, 5, -1];$
	$>>B = [6, -4, 27]';$
	$>>inv(A) * B$
	$ans =$
	5.0000
Division between matrices is not defined in linear algebra, yet MATLAB® can use inverse multiplication, or left division (note that the second version is more optimized).	3.0000
	-2.0000
	$>>A \backslash B$
	$ans =$
	5
	3
	-2
	$>>A = [1, 2, 3; 4, 5, 6; 7, 8, 9]; s = 2;$
	$>>A\string^s$
Power of matrix against a scalar can only be done for a square matrix	$ans =$
	30 36 42
	66 81 96
	102 126 150

Although the above is true, there are certain cases where it is required to multiply element by element or to power each of the elements by a scalar. For those cases, MATLAB has implemented element-wise operations using a dot before the mathematical symbol (.* for multiplication and .^ for power). For example,

$>> A = [1, 2, 3; 4, 5, 6; 7, 8, 9];$
$>> A. * A$

$ans =$

$$\begin{array}{ccc} 1 & 4 & 9 \\ 16 & 25 & 36 \\ 49 & 64 & 81 \end{array}$$

$ans =$

$>> A.\hat{}A$

$$\begin{array}{ccc} 1 & 8 & 27 \\ 64 & 125 & 216 \\ 343 & 512 & 729 \end{array}$$

2.1.5 Basic Commands clc, who, clear, save, and load

As previously stated, using semicolon stops the output from being shown. Yet, in some cases, the command window gets overcrowded with variables that are no longer easy to follow. Hence, the function of the command "clc" is to clean the command window without affecting the workspace. This translates in a clean space without affecting any of the variables.

There are some commands that are helpful while declaring variables. The command "who" shows all the variables currently in the workspace; similarly, the command "who" will show all the variables and display its size, bytes, class, and attributes. This is similar to what is displayed on the workspace window.

Another important command is "clear". This command without any arguments removes all variables

$>> clear$

Similarly, "clear" erases all the variables. For example,

$>> clear \quad c$

will remove the variable c from the workspace.

It is important to realize that after closing MATLAB, it will erase all variables, but in some cases, it is required to use that variable in another session. Hence, the command "save" will store a variable and the command "load" will make a variable saved in a previous session available. For example, $>> save('test.mat','d')$ will save the variable d in a file called "test.mat" and $>> load('test.mat')$.

These functions and some other relevant functions are listed below:

Command	Definition
clc	cleans the command window.
who	shows a list of the variables in the workspace.
clear	erases the variables from the workspace.
save	saves the variables from the workspace into a *.mat file
load	loads the variables to the workspace from a *.mat file.
close	closes all open figures.
format	defines the format of the numerical values that are presented.
help	opens the help section.
quit or exit	closes MATLAB.

2.1.6 Basic Functions

Besides the aforementioned commands, MATLAB has a large number of incorporated functions. MATLAB's main syntax for any function of n input variables $x_1, \ldots, x_n$ and m output variables $y_1, \ldots, y_m$ is as follows:

$$[y_1, ..., y_m] = function(x_1, ..., x_n)$$

Some of MATLAB's incorporated elementary mathematical functions are shown in Table 2.2.

TABLE 2.2 Elementary Mathematical Functions in MATLAB®

Function	Definition
	Matrix Creation
zeros()	creates array of all zeros.
ones()	creates array of all ones.
eye()	creates array of all zeros and the main diagonal filled with ones.
magic()	matrix constructed from the integers 1 through n2 with equal row and column sums.

(Continued)

TABLE 2.2 (*Continued*) Elementary Mathematical Functions in MATLAB®

Function	Definition
	Simple Operation
sum()	gets the sum of all the elements in an array.
prod()	gets the product of all the elements in an array.
abs()	gets the absolute value of each element in an array.
	Statistics Operation
min()	gets the minimum of all the elements in an array.
max()	gets the maximum of all the elements in an array.
mean()	gets the mean of all the elements in an array.
std()	gets the standard deviation of all the elements in an array.
	Rounding
round()	gets the round value of each element in an array.
fix()	gets the round number to zero of each element in an array.
floor()	gets the round number to the nearest integer lower than or equal to each element in an array.
ceil()	gets the round number to the nearest integer greater than or equal to each element in an array.
	Power and Exponentiation
power()	gets the power to a certain value of each element in an array.
sqrt()	gets the square root of each element in an array.
exp()	gets the exponential of each element in an array.
log()	gets the natural logarithm of each element in an array.
	Trigonometric Functions (Radians)
sin()	gets the sine of each element in an array.
cos()	gets the cosine of each element in an array.
tan()	gets the tangent of each element in an array.
asin(),	gets the arcsine or arccosine of each element in an array.
acos()	gets the arctangent (between $-\pi$ and π) of each element
atan2()	in an array.
	Complex Numbers
complex()	creates a complex number.
abs()	gets the modulus of a complex number.
angle()	gets the phase angle of the complex number.
real()	gets the real part of a complex number.
imag()	gets the imaginary part of a complex number.

2.1.7 Programming

At this point, all the commands have been written directly at the command window; however, that is not always practical, especially if several lines of code need to be repeated over and over. Hence, MATLAB has a programming environment where several lines of code can be created and run multiple times. MATLAB programming is carried out by means of simple text files with *.m extensions, which contain MATLAB commands. To use files, it requires the following steps:

- Edit a file with a text editor or the MATLAB editor.

- Save the file with a *.m extension.

- Specify the path of the file to MATLAB.

- Execute the program by typing its name in the command window, or if the file is opened within MATLAB, press the run button or press $F5$.

The MATLAB editor can be opened by pressing the New Script button in the Ribbon (see Figure 2.1 upper left corner) or by pressing ctrl + n, where commands can be written one after the other. This editor is normally docked between the current folder address and the command window (Figure 2.4).

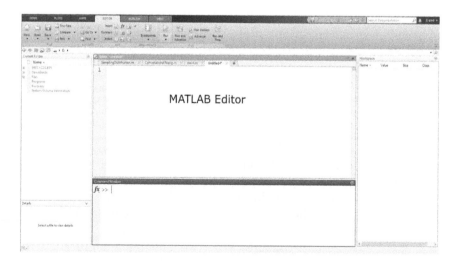

Figure 2.4 MATLAB® editor.

There are two types of files: the scripts and the functions. A script is formed of a sequence of MATLAB commands, yet the scripts can only operate on the variables that are hard-coded into their m-file. On the other hand, functions have input and output parameters. Scripts are more useful when the data does not change, while functions are mostly general-purpose program and can be applied to different data.

MATLAB, unless otherwise indicated, works sequentially, which means that it will run one line of code, then the next, and so on. For instance, a simple script with the following lines at the end will result in the variable m having the value of 5 (note that the last line has no semicolon).

```
m = 1;
m = 'c';
m = 5
```

An important part of every programming language are comments, which are the notes containing a certain instruction or set of instructions for the execution of program. In MATLAB, comments are written using the percentage sign %. In many cases, it is required to comment or uncomment quickly more than one line of code, which can be done by pressing the hotkey "ctrl + r" or "ctrl + t", respectively. For the previous example, for suppressing the last line and ending with m with a character value $'c'$, comment will be made for the last line of the code.

```
m = 1;
m = 'c';
% m = 5;
```

Custom functions have syntax similar to basic functions. For instance, a function of n input variables $x_1, \ldots, x_n$ and m output variables $y_1, \ldots, y_m$ is called as:

$$[y_1, ..., y_m] = function(x_1, ..., x_n)$$

It is recommended that the function has the same name as the .m file.

A simple example of a function will be a program that returns the addition and subtraction of two variables.

```
function [addition, subtraction] = addsub(A, B)
% This program calculates the addition and
         subtraction of two variables
% Input data: Two variables or matrices A and B.
```

```
% Output data: Two variables or matrices:
       addition = A + B, subtraction = A - B.

addition = A + B;
subtraction = A - B;
end
```

This program needs to be saved with the name "addsub.m", and to call it, it just requires to type

$$>> [M, N] = addsub(eye(3), \quad eye(3))$$
$M =$

$$\begin{array}{ccc} 2 & 0 & 0 \\ 0 & 2 & 0 \\ 0 & 0 & 2 \end{array}$$

$N =$

$$\begin{array}{ccc} 0 & 0 & 0 \\ 0 & 0 & 0 \\ 0 & 0 & 0 \end{array}$$

It is important to observe the first commented lines that explain how the code works. This is especially useful when using the command "help sumres", which will return the initial commented lines.

2.1.8 Conditionals

In some cases, it is important to limit the execution of a certain code unless a determined condition is satisfied, hence the importance of conditionals. Before moving forward, the relational and logical operators are listed in the following tables:

Relational Operators	
Operator	**Definition**
>	greater than.
>=	greater than or equal to.
<	less than.
<=	less than or equal to.
==	equal to.

Logical Operators	
Operator	**Definition**
&&	logic and.
||	logic or.
~	logic not.

For the control structure, there exists the **if** and the **switch** statements.

2.1.8.1 IF

The general structure of the **if** statement is

```
if expr1
    statements
elseif expr2
    statements
    .
    .
    .
else
    statements
end
```

For example, suppose you wanted to divide one number by another number, you might first check to see that the divisor is not equal to 0.

```
a = 3;
b = 0;
% b is not equal to 0
if (b ~= 0)
    c = a/b;
else
    disp('Cannot␣divide␣by␣0');
end
```

Another example shows the calculation of y where $y = -1$ when $x < 0$, $y = 2$ when $x > 2$, and $y = 0$ for any other value of x.

```
if x < 0
    y = -1
elseif x > 2
    y = 2
else
    y = 0
end
```

2.1.8.2 Switch

In some cases, there are many conditions that the **if** has to evaluate. For those cases, there exists a more suitable condition called switch that evaluates the variable for many situations. The syntax for the use of the **switch** statement is as follows:

```
switch variable
   case value1
      statements
   case value2
      statements
   case value3
      statements

      .

      .

      .

   otherwise
      statements
end
```

In this structure, the instructions are executed based on the block in which the variable is equal to the value $(value1, value2, \ldots, valorn)$ of any of the cases. When the value of the structure does not correspond to any case, the instructions found in otherwise are executed. For example, the switch will print the name of the number if it is found within that variable.

```
a = 1;
switch a
   case 1
      disp('one')
   case 1
      disp('one')
   case 2
      disp(two)
   otherwise
      disp('any other number')
end
```

Note that in this example, there are two cases with the same value. In case of similar values, the first one found will be the one to be executed.

2.1.9 Loops

There are two types of loops: the "for" loop and the "while" loop.

2.1.9.1 For Loop

A **for** loop is a set of operations that are going to be executed for a known specific number of times.

```
for index = values
    <program statements>
    ...
end
```

Values can be one either initval:endval that increments the index variable from initval to endval in steps of 1, or initval:step:endval that increments index by the value step on each iteration (or decrements when the step is negative) and repeats the execution of program statements until index is greater than endval.

For example, to form a vector $v = (1, 2, \ldots, n)$, it can be created by iterating from i till it reaches n (note that this can be done more efficiently like it was done in Section 2.1.4). In most cases, it is normally preferred to initialize the array to save some execution time.

```
n=100;
v=zeros(1,n);
for i=1:n
        v(i)=i;
end
```

It is also possible to use an array and loop through its values

```
for a = [37, 35, 18, 21, 43]
    disp(a)
end
```

In MATLAB, it is possible to use one loop inside another loop. The syntax for a nested **for** loop statement is as follows:

```
for m = 1:j
    for n = 1:k
        <statements>;
    end
end
```

For example, two **for** loops are running: one for i running from 1 to 5 in steps of 1 and the other nested inside the first for j running from 2 to 8 in steps of 2. In the inside nested loop, it calculates the sum of the current i and j values and finally outputs the sum

```
for i = 1:5
    for j = 2:2:8
        b(i,j) = i + j;
```

```
      end
end
disp(b)
```

2.1.9.2 While Loop

The **while** loop executes statements repeatedly as long as a specified condition is true. The syntax for a **while** loop statement in MATLAB is as follows:

```
while <expression1>
    <statements>
end
```

An expression is considered true when the result is nonempty and only contains nonzero elements (logical or real numeric). Otherwise, the expression is considered false. For example, a code starts with a value of 5 and stops once the value reaches 10.

```
a = 5;
while( a < 10 )
    disp(['value␣of␣a:', num2str(a)])
    a = a + 1;
end
```

2.1.9.3 Break and Continue Loops

There are two statements that are useful for the execution of the loops: the **break** and the **continue**. The **break** terminates the execution of the **for** or **while** loop, and the **continue** passes the control to the next iteration of a **for** or **while** loop. For both cases, the statements that appear after the break or continue statements are not going to be executed (see Figure 2.5).

For example, a while loop end after 20 iterations is affected by the break condition that ends the while loop when a value bigger than 15 is reached.

```
a = 10;
while (a < 20 )
    disp(['value␣of␣a:' num2str(a)])
    a = a + 1;
    if( a > 15)
        % terminate the loop using break
        break;
    end
end
```

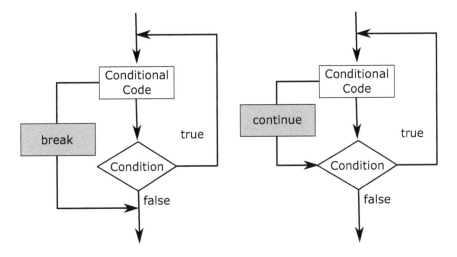

Figure 2.5 Break and continue executions.

In nested loops, break exists only from the loop in which it occurs. Control passes to the statement following the end of that loop. For example, nested loops are used to display all the prime numbers from 1 to 100, which breaks every iteration.

```
for  i=2:100
    for  j=2:100
        if(~mod(i,j))  % factor  found ,  not  prime
            break;
        end
    end
    if(j > (i/j))
        disp([num2str(i)  'Lis prime'])
    end
end
```

The continue statement in MATLAB works somewhat like the break statement, but, instead of forcing termination, **continue** forces the next iteration of the loop start, skipping any code in between. For example, the following code skips any iteration between the numbers 15 and 18.

```
a  =  10;
aa  =  0;
while  a  <  20
```

```
a = a + 1;
if a > 15 && a < 18
    % skip the iteration
    aa = 0;
    continue;
end
disp(['value␣of␣a:␣', num2str(a)])
aa = a^2;
disp(['Its␣square␣value␣is:␣', num2str(aa)])
end
```

2.1.10 Graphs

MATLAB contains several functions to make a high-quality, two-dimensional or three-dimensional graphs. The most common function to make a two-dimensional graph is the plot function. The argument of the function must have the pair of variables that are going to be graphed. For example, the following variable shows the use of the plot function to make a graph of the cosine for different values for an angle from 0 to 2π (Figure 2.6).

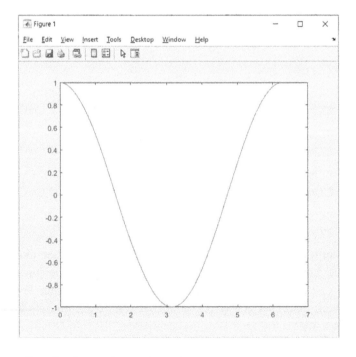

Figure 2.6 Graph of a cosine using plot.

```
t = 0 : 0.1 : 2*pi;
c = cos(t);
plot(t, c)
```

On the other hand, the plot function can have extra inputs that configure the color and style of a line, which has to be included between quotation marks after the pair of variables at the function. This is exemplified for the previous example using red and an asterisk in each of the points (Figure 2.7):

```
t = 0 : 0.1 : 2*pi;
c = cos(t);
plot(t, c, 'r*')
```

Some of the most common styles are included in Table 2.3.

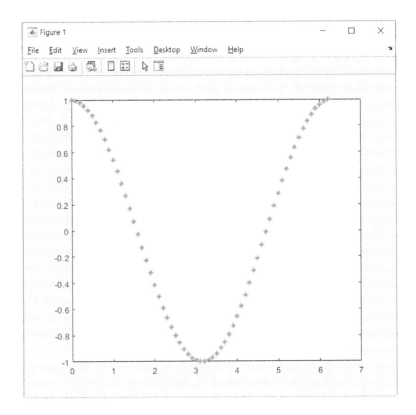

Figure 2.7 Graph of a cosine using plot with color and line properties.

TABLE 2.3   Common Styles used in MATLAB®

Symbol	Color	Symbol	Line Style	Symbol	Marker
k	Black	—	Solid	+	Plus sign
r	Red	— —	Dashed	o	Circle
b	Blue	:	Dotted	*	Asterisk
g	Green	−.	Dash-dot	.	Point
c	Cyan	none	No line	x	Cross
m	Magenta			s	Square
y	Yellow			d	Diamond

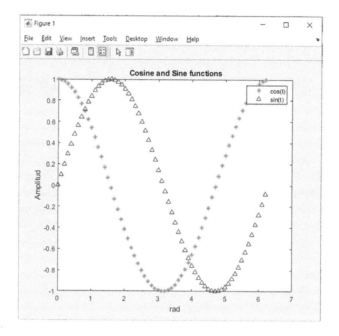

Figure 2.8   Graph of a cosine and sine using plot with color, line properties, title, labels, and legends.

To plot a second figure, the **hold on** function can be used. Also, to add some title, labels to the axis, and legend to the plots, the **title()**, **xlabel()** and **ylabel()**, and **legend()** functions can be used. Using the previous example, a second plot is used with the sine function added and the respective title, labels, and legends (Figure 2.8).

```
t = 0 : 0.1 : 2*pi;
c = cos(t);
s = sin(t);
plot(t, c, 'r*')
```

```
hold on
plot(t, s, 'b^')
title('Cosine␣and␣Sine␣functions')
xlabel('rad')
ylabel('Amplitud')
legend('cos(t)','sin(t)')
```

To create three-dimensional graphs, a standard function is the **surf()** function. The **surf()** function can be used in tandem with the **meshgrid()** function, which generates the values for the variables for a certain interval. For example, if the function $z = x^2 + y2$ needs to be graphed, the following commands can be used (Figure 2.9):

```
[x, y] = meshgrid(-1:0.1:1, -2:0.2:2);
z = x.^2 + y.^2;
surf(x, y, z)
```

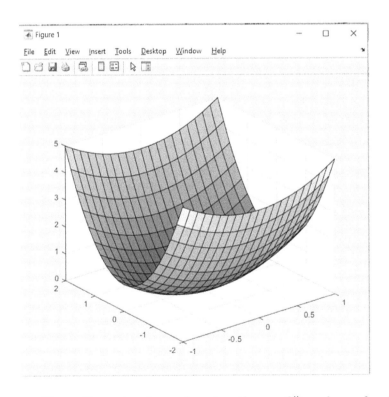

Figure 2.9 Three-dimensional graph using the **surf()** and **meshgrid()** functions.

2.1.11 Examples

Example 2.1 *Program to calculate the maximum of four numbers*

```
maxout = maxfive(3, 6, 8, 5);
disp(maxout)

function out = maxfive(n1, n2, n3, n4)
   % Calculates the maximum of four numbers
   out = n1;
   if(n2 > out)
      out = n2;
   end
   if(n3 > out)
      out = n3;
   end
   if(n4 > out)
      out = n4;
   end
end
```

Solution 2.1 *8*

Example 2.2 *Program to calculate the maximum of an array*

```
arr = [3, 6, 8, 5, 4];
maxout = maxarr(arr);
disp(maxout)

function out = maxarr(in)
   % Calculates the maximum of an array numbers
   sizearr = size(in, 2);
   out = 0;
   for n = 1:sizearr
      if(in(n) > out)
         out = in(n);
      end
   end
end
```

Solution 2.2 *8*

Example 2.3 *Program to determine the number of years to double an investment*

```
rate = 4.5;
init = 125000;
target = 2*init;

balance = init;
year = 0;

while balance < target
    year = year + 1;
    interest = balance*rate/100;
    balance = balance + interest;
end

disp(['the target will be reached in '
    num2str(year) ' years'])
```

Solution 2.3 *The target will be reached in 16 years*

Example 2.4 *Program to plot a sine continuously*

```
T = 10;
t = 0.0;
dt = 0.01;

while t < T
    t = t + dt;
    y = sin(2*pi*t);
    plot(t, y, 'r*'), hold on
    pause(0.01)

end
```

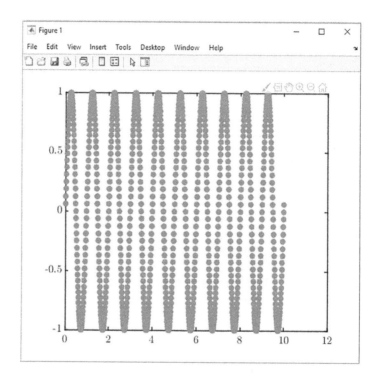

Solution 2.4

Example 2.5 *Program to sort an array (bubble sort)*

```
arr = [3, 6, 8, 5, 4];
swapped = false;
n = size(arr, 2);
while ~swapped
    swapped = true;
    for m = 1 : n - 1
        if arr(m) > arr(m + 1)
            swapped = false;
            temp = arr(m);
            arr(m) = arr(m+1);
            arr(m+1) = temp;
        end
    end
end
disp(arr)
```

Solution 2.5 *3 4 5 6 8*

2.2 SIMULINK FUNDAMENTALS

Simulink® is a graphical environment for model-based simulation of dynamic systems. Some main advantage of Simulink is its ability to model nonlinear systems where the transfer function may not be easily solvable and can take initial conditions to the system.

The models created in Simulink contain blocks and signals (which can be labeled) that can be placed on a background:

- **Blocks** are the functions that can have varying inputs and outputs. These inputs and outputs can be either constant or dynamic.

- **Signals** are the lines that connect and transfer the information between the blocks.

Figure 2.10 shows a simple Simulink model that contains both blocks and signals.

Simulink can be open either by typing *simulink* in the command window or by clicking on the Simulink icon on the home ribbon (see Figure 2.11).

At this point, it opens Simulink Start Page where several features can be selected (see Figure 2.12a). For this brief introduction, only the blank model is important (the other options are more to create either a more elaborate project or a collaborative work). Once open, an upper

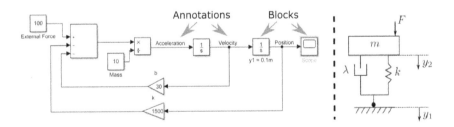

Figure 2.10 A Mass-Spring-Damper system modeled within Simulink®.

Figure 2.11 Simulink® icon.

Figure 2.12 (a and b) Opening Simulink® and its toolbar.

toolbar exists with four main features: File Manager, Model Settings, Run Model, and Build Model.

File Manager allows us to create a new model, to open an existing model, or to save the current model. Model Settings has the Library Browser where many features are saved, the model parameters (Cog icon) allow us to change how the model is going to be numerically solved, and the model explorer has most of the information of all the blocks saved. Run Model has three main features: the run keys that start, stop, step, or restart the model; the time in seconds the model will be executed (center box); and the different modes (e.g., normal or hardware). Finally, Build Model helps in interacting with hardware and will build the model before it can run. All these features are shown in Figure 2.12b.

2.2.1 Working with Blocks

There are mainly two ways to add blocks to a model: Quick Search and Library Browser. Quick Search looks for blocks by clicking on the background of model canvas, typing in a search term, and selecting the block from the result. Library Browser is located on the Model Settings and shows all blocks available in Simulink, sorted by folders (e.g., "Math Operations" or "Signal Routing"). Also, there is a search bar on the top left. Once the block is found, it can be added by

- Sources that are used to generate various signals such as sinusoidal, step, and constant.

- Sinks that are used to output or display signals, for instance, using a scope, XY graph, and workspace.

- Continuous that has continuous-time system elements, such as transfer functions and state-space models.

- Discrete that has linear, discrete-time system elements, such as unit delay, discrete transfer functions, and discrete state-space models.

- Math operations that contain some of the most common math operations sum, product, multiplication, gain, and absolute value.

- Ports and subsystems that have some useful blocks to build a system, for example, sysbsystem, enable port, and inputs/outputs(in1 and out1).

- User-defined functions that implement custom functions in Matlab.

- Lookup tables that have functions defined as discrete data, for instance, 1-D Lookup Table.

- Logical and bit operations that have boolean operators for comparisons, compared to zero and logical operators.

- Signal routing that helps in organizing how the signals are going to move from blocks, for instance, mux, BusCreator, Goto, and Switch.

2.2.2 Block Settings

All the blocks have their own settings that can be defined in the Block Parameters and the Block Properties.

Block Parameters are opened by double clicking the block, where you can change the settings for that specific block. For example, in the integrator, block has many parameters such as the initial condition or absolute tolerance (see Figure 2.13).

Block Properties can be opened using right-click, where you can select *Properties*. This contains the general settings on how the block will behave as part of the larger model. For example, it can be used to provide a description of the block, change the information that is displayed about the block in the Block Annotation tab, or use callbacks in the Callbacks tab.

2.2.3 Simulink Models and MATLAB Variables

MATLAB variables can be used by Simulink blocks. These variables have to be defined in the MATLAB workspace before the Simulink model run;

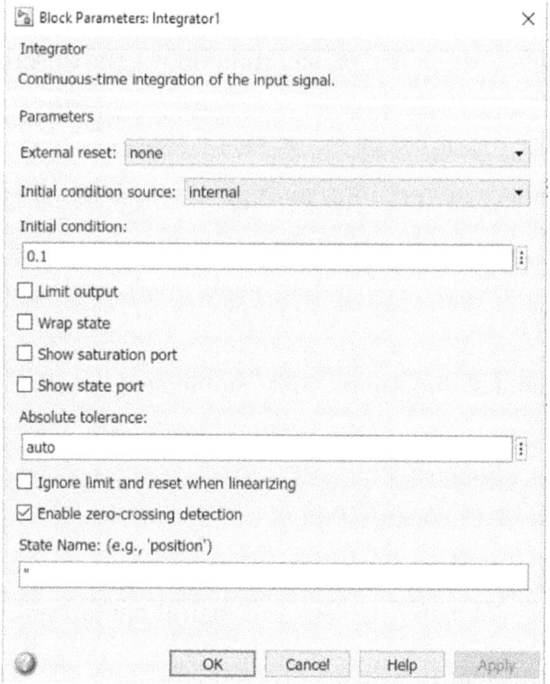

Figure 2.13 Integrator block properties.

otherwise, an error appears stating there is an undefined function or variable. It is common to define several variables in an m-script as initial conditions in a Simulink program; in case some of them have to be changed in the Simulink program, opening and changing them in the m-script and run it and all the block parameters in Simulink with those variables will be updated.

2.2.4 Simple Simulink Example

A simple Simulink example is the one displayed at the beginning of this section. That is a Mass-Spring-Damper system with an extra force applied. The system is a second-order differential equation that has a spring constant $k = 1500.0\,N/m$, a damping constant $b = 30.0\,Ns/m$, a mass of the device $m = 10.0\,kg$, an input force $F = 0.0, N$, and an initial condition of position and velocity of $0.1\,m$ and $0.0\,m/s$ (see Figure 2.14).

This model was simulated for 10 seconds, and the result was displayed using a source. This can be observed in Figure 2.15.

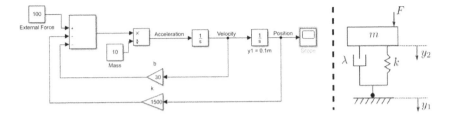

Figure 2.14 Mass-Spring-Damper system.

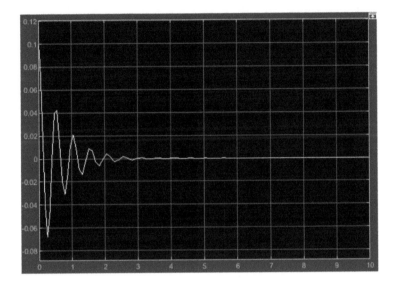

Figure 2.15 Mass-Spring-Damper system result after it was simulated for 10 seconds in Simulink®.

2.3   GENERAL INTRODUCTION TO LabVIEW

National Instruments is a company that develops hardware and software which allow generating academic and industrial products [54]. One of the best industrial and academic software packages developed by National Instruments for creating software that can be connected with hardware is called LabVIEW. This software can be connected directly with hardware such as DAQ, My DAQ, and CompactRIO [56].

This software can be used by academia and industry for proposing, testing, and validating new methodologies as well as designing rapid prototyping and moving it to a final product using the same software and hardware. Thus, this chapter is a brief introduction to LabVIEW. If further information regarding LabVIEW is required, it is recommended to review a specialized book of LabVIEW [5,53,79]. NI community is also an excellent resource for finding examples and information about LabVIEW.

LabVIEW is a graphics programming that has two main windows: the block diagram window and the front panel window, as shown in Figure 2.16.

The starting window provides information about the LabVIEW version and examples that could be useful for starting programming (see Figure 2.17). Since LabVIEW was created for emulating real instruments by virtual instruments, the main program has an extension VI that means virtual instruments.

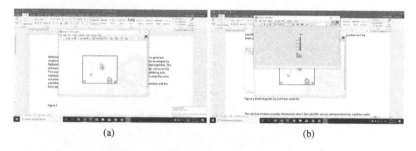

(a)   (b)

Figure 2.16 Block diagram (a) and front panel (b).

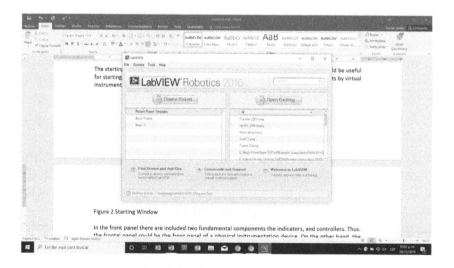

Figure 2.17 Starting window.

In the front panel, there are two fundamental components: the indicators and controllers. Thus, the frontal panel can be the front panel of a physical instrumentation device. On the other hand, the block diagram has all the blocks and nodes to build a LabVIEW program. The block diagram menu and frontal panel menu are illustrated in Figure 2.18.

The main elements in the front panel that are considered in the program code are the run button, the run continuously button, the stop button, and the pause button. These elements are in the toolbar. There is an important button called the highlight execution button, which is defined by a lamp icon.

For building a basic VI program that can add two variables, there is a specific toolkit in the block diagram menu so it has to be selected the specific toolkit and look for the block that is required (addition) in the programming menu inside the numeric option. When an input variable is selected, it can be integer, chart, float, or fix, and it appears in the block diagram window and frontal panel at the same time. Thus, you can change the variable's value when the VI program is running, as shown in Figure 2.19.

(a)   (b)

Figure 2.18 The front panel menu (a) the block diagram menu (b).

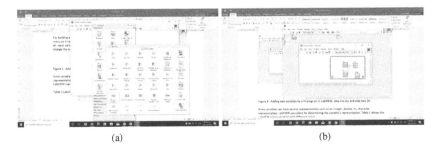

(a)   (b)

Figure 2.19 Adding two variables by a VI program in LabVIEW, step one (a) and step two (b).

Since variables can have several representations such as an integer, double, fix, and character representation, LabVIEW uses colors for determining the variable's representation. Figure 2.20 shows how LabVIEW uses different colors according with the representation (integer, double, fixed point, array, etc).

LabVIEW has also control blocks such as while, for, and case among others. These control blocks are defined based on the general structure of a conventional program. For instance, the "while" loop requires a stop condition instead of using the number of iterations as a "for" block does. Figures 2.21–2.23 give a general description when those control blocks are programming on LabVIEW.

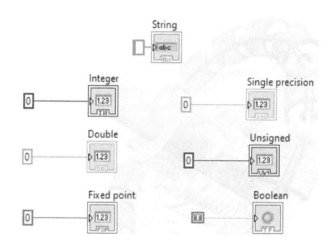

Figure 2.20 LabVIEW representation.

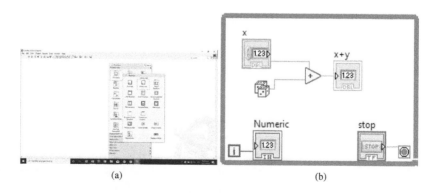

Figure 2.21 Path for finding functional blocks (a) and the while functional block (b).

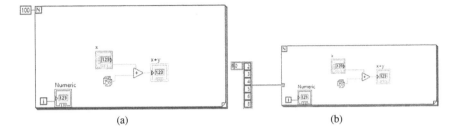

(a) (b)

Figure 2.22 The functional block "for" that uses the iteration number as stopping condition or uses the number of elements in the array to index the "for" as stopping condition.

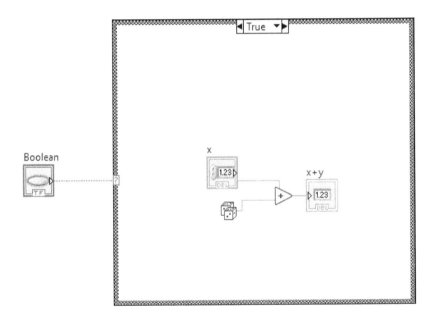

Figure 2.23 Case functional block using a boolean condition.

Figure 2.24 depicts how a while, case, and for can be implemented in a basic program by the previous blocks presented above.

Besides, LabVIEW has a parallel structure, which means LabVIEW can run in parallel with the blocks if they are not directly linked. So if the program requires a serial execution, it is important to create dependencies as shown in Figure 2.25.

There are two basic options to plot points in LabVIEW: A chart helps to plot point by point, and the graph helps to plot an array. Figure 2.26

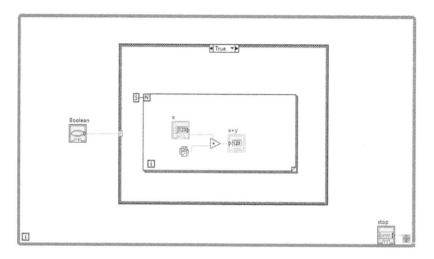

Figure 2.24 Basic program using for, while, and case structures.

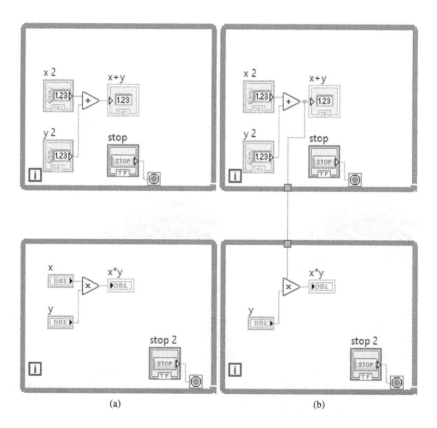

(a)   (b)

Figure 2.25 Parallel execution (a) and serial execution (b).

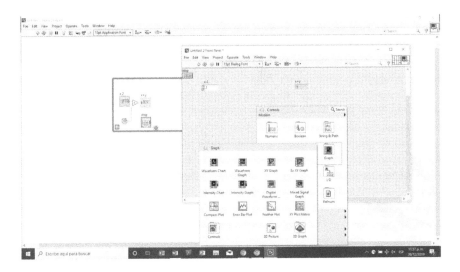

Figure 2.26 Path for getting the chart and graph plots.

illustrates the path for getting the chart and graph plot in the front panel. Figures 2.27 and 2.28 show the chart and graph options.

The arrays in LabVIEW have to contain the same type of elements and representation; for instance, an array of integer numbers is illustrated in Figure 2.29. A cluster can include different elements as shown in Figure 2.30, which has a cluster of double, boolean, and string elements. As a result, also an array of clusters can be constructed, as shown in Figure 2.31. The data flow in LabVIEW is illustrated in the example regarding the calculation of an average value of four numbers, which is presented in Figure 2.31.

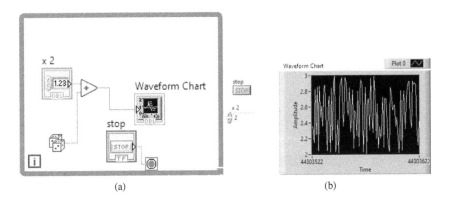

(a)   (b)

Figure 2.27 (a and b) Chart plot using LabVIEW.

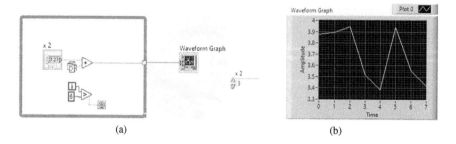

(a)

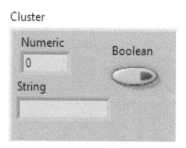

(b)

Figure 2.28 (a and b) Graph plot using LabVIEW.

Figure 2.29 Array of integers.

Figure 2.30 Cluster with several elements of different type.

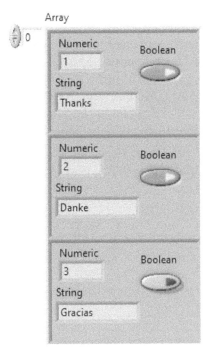

Figure 2.31 Array of clusters.

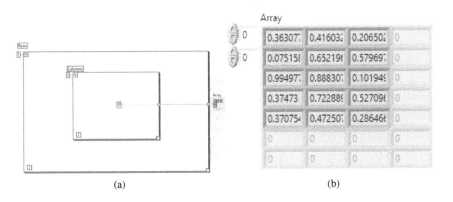

(a)   (b)

Figure 2.32 (a and b) For loops to create a matrix.

One way to generate a matrix using "for" loops is a nest configuration, in which the inner loop is defined by a "for" structure and the outer loop by another "for" loop. Figure 2.32 shows the configuration of the "for" loops.

LabVIEW operations with matrices are friendly, and it is possible to add, subtract, and multiply matrices, as shown in Figure 2.33.

There are special blocks for using Matlab code into LabVIEW so this code can be run using specific libraries that are supported by LabVIEW. Figure 2.34 depicts the block "for" using Matlab code by the MathScript RT module.

Besides, LabVIEW has special toolkits like simulation toolkit, in which the user can build a model using linear and nonlinear systems. For instance, if a second-order system is modeled using transfer functions

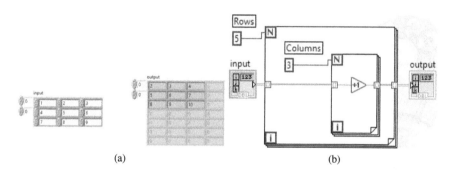

(a)   (b)

Figure 2.33 (a and b) Adding operation for a matrix.

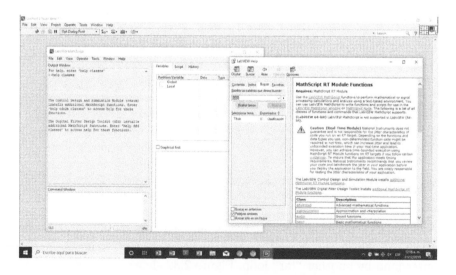

Figure 2.34 MathScript documentation and window into the toolbar of the front panel.

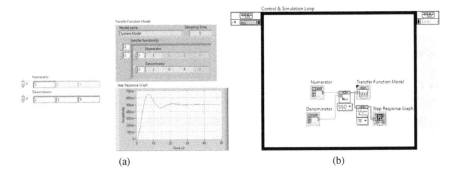

Figure 2.35 (a and b) Simulation of a second-order linear system using the simulation toolkit.

and simulated when a step input is applied, a graphical representation is shown in Figure 2.35.

It is important to remark that a VI program can be directly connected with hardware like a DAQ [6] so the program code can read data from external sources as well as write data. Thus, the program can interact with external variables from physical systems. Hence, metaheuristic optimization can run online when that hardware is interconnected with the VI program.

II

Metaheuristic Optimization

CHAPTER 3

Basic Metaheuristic Optimization Algorithms

3.1 EXHAUSTIVE SEARCH

The most basic form of search is to look at all possible solutions and to decide which one is the best. This method is called exhaustive search or brute force method. Although this search ideally guarantees the finding of an optimal solution, the time it requires is high, and in some cases, it is inefficient. As an alternative, it is possible to use an algorithm that limits the search using some kind of heuristics.

3.2 RANDOM OPTIMIZATION

A simple way of searching the space is random optimization. The method basically searches the space using a random number over certain number of iterations, and if at any step a better solution is found, then it replaces the current one. Although this method does not guarantee the finding of the optimal solution, with enough iterations it may find a solution that is good enough. The pseudocode for random search is shown in Figure 3.1.

3.3 NELDER–MEAD ALGORITHM

The Nelder–Mead algorithm [51,75] is a blackbox optimization method that operates similar to some evolutionary algorithms. Differently than some other algorithms, the Nelder–Mead algorithm does not solely resort to selection for improving the average solution, and it does not contain stochastic decision.

```
1: Set Objective Function f(x), x = [x₁,..., xₐ].
2: Initialize position x̂
3: while t =< Max iterations do
4:     Create a new solution randomly using x̂ .
5:     if f(x*) > f(x̂),
6:         x̂ = x* then
7:     end if
8:     Select current best x̂
9: end while
```

Figure 3.1 Random optimization pseudocode.

The method determines from the three vertices what is the worst vertex, which would be the one with the largest function value, and this worst vertex will be replaced with a new vertex to form a new triangle. This process will be repeated, thus having different shapes of triangle, and in general, this method is trying to make the triangle smaller until the minimum is found. The algorithm has mainly 4 steps:

- Reflection

- Expansion

- Contraction

- Shrinking.

The Nelder–Mead method starts with a triangle with three vertices x_1, x_2, and x_3. These vertices are evaluated in $f(x)$, and the order of selecting the smaller is the best, the second smaller is good and the largest is the worst, which are defined as B, G, and W, respectively.

After the vertices are ordered, a midpoint between B and G is calculated in order to find a new point that will replace W. The midpoint is the average of the coordinates:

$$M = \frac{B+G}{2} \tag{3.1}$$

Then, since the function decreases from W to either B or G, it is possible that a smaller value can be found at the opposite side of W that passes through the midpoint M. Hence, a **reflection** point can be calculated by getting the distance d from W to M and using this distance on the other side. The formula for R is

$$R = M + (M - W) \tag{3.2}$$

If $f(R)$ is smaller than $f(W)$, then the movement is assumed as correct. At this point, it can be supposed that the minimum is farther from R, so the segment is **expanded** to the point E by moving an extra distance d. The expansion point E can be calculated as follows:

$$E = R + (R - M) \tag{3.3}$$

Assuming that $f(E)$ is smaller than $f(R)$, E will be a better vertex.

If the function $f(R)$ is larger than or the same as $f(W)$, then another point must be calculated. At this point, two **contraction** midpoints C_1 and C_2 are selected in between the lines WM and RM, respectively. From these two points, the one with the smallest function value will set as C and it will replace W.

In the case none of the previous steps work, then the triangle is **shrunk** toward B. The step replaces G with M and W with S (midpoint between W and B). The four steps are illustrated in Figure 3.2 and their pseudocodes in Figure 3.3.

Some variables can be included, such as the reflection coefficient α, the expansion coefficient γ, the contraction coefficient ρ, and the shrinking coefficient σ.

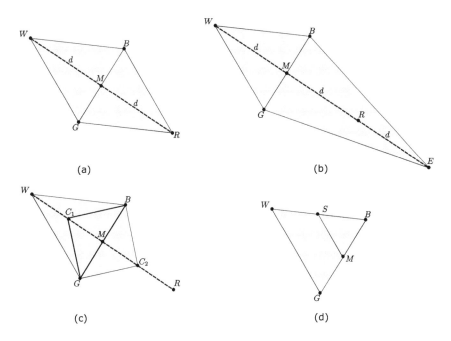

Figure 3.2 The four steps of the Nelder–Mead algorithm.

IF $f(R) < f(G)$, THEN Perform Case (i) {either reflect or extend} ELSE Perform Case (ii) {either contract or shrink}

BEGIN {$Case(i)$.}
if then$f(B) < f(R)$ **THEN**
  replace W with R
else
  Compute E and $f(E)$
  if then$f(E) < f(B)$ **THEN**
   replace W with E
  else
   replace W with R
  end if
end if
END {$Case(i)$.}

BEGIN {$Case(ii)$.}
if then$f(R) < f(W)$ **THEN**
  replace W with R
end if
Compute $C = (W + M)/2$
or $C = (M + R)/2$ and $f(C)$
if then$f(C) < f(W)$ **THEN**
  replace W with C
else
  Compute S and $f(S)$
  replace W with S
  replace G with M
end if
END {$Case(ii)$.}

Figure 3.3 Nelder–Mead algorithm pseudocode.

CHAPTER 4

Evolution Algorithms

4.1 GENETIC ALGORITHMS

One of the most popular methods for optimization is the genetic algorithm (GA), which was developed by John Holland and his students [32]. This method uses the principles of natural selection to find the objective optimum within an objective function [32]. Furthermore, GA uses the principles of evolution of species as it considers keeping characteristics that have an advantage over the rest. The algorithm also considers the process of reproduction, crossing individual, and the idea of mutation that help in creating new characteristics that might be beneficial.

In particular, the process has an initial population, and it is evaluated using the objective function. Then, based on the evaluation, individuals with highest value are more likely to be selected for reproduction. Afterward, the selected individuals are crossed (or mated) to create a new generation. As an additional step, mutation randomly deform the chains to create new features.

The previously described process is repeated to create new generations that might produce better individuals by keeping beneficial characteristics, mixing them, and mutating to produce unseen features. Since this is also an iterative process, after several generations an optimal value is likely to be found.

Similarly, mathematically GA starts creating individuals with different features. Then, the features of the best individuals are crossed and some individuals get mutated to create a new set with new characteristics. Afterward, the process repeats over several iterations, which might result in the end in an individual with better characteristics. An example

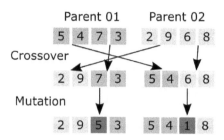

Figure 4.1 GA next generation.

1: Definition of objective function $f(x)$
2: Set Population size n
3: Set percentage of crossover and mutation
4: Number of iterations t
5: Initialization of population X_i
6: **while** $t =< Max\ iterations$ **do**
7:    Calculate the fitness to all the population f_i
8:    Select best solutions
9:    Create New generation
10:    Perform Crossover
11:    Perform Mutation
12: **end while**
13: Set best individual as optimum

Figure 4.2 Genetic algorithm pseudocode.

of crossing and mutation processes is shown in Figure 4.1, where from two parents with four features each, two are crossed to generate new individuals, and later, these two offspring have one feature mutated.

The GA pseudocode is shown in Figure 4.2.

4.2 SIMULATED ANNEALING

Simulated annealing (SA) [17,40] is an optimization algorithm that is based on metallurgy and how metals cool down. In metallurgy, annealing is a process where the metal cools down slowly to avoid imperfections. Annealing process is carried out by heating the metal and cooling it down slowly, thus permitting atoms to arrange into a minimal energy state.

Mainly, the annealing process is done by two steps: increasing the temperature of the heat bath until the metal melts and decreasing the

1: Set Objective Function $f(x)$, $x = [x_1, \ldots, x_d]$.
2: Initialize system configuration
3: Initialize the first solution x_0
4: Initialize T with a large value
5: **while** $t =< Max\ iterations$ **do**
6:   **repeat**
7:     Apply random permutation Δx to the state: $x_{i+1} = x_i + \alpha \Delta x$
8:     Evaluate $\Delta E(x) = E(x + \Delta x) - E(x)$:
9:     **if** $\Delta E(x) < 0$ **then**
10:       keep the new state,
11:     **else**
12:       accept the new state with probability $P = e^{-\frac{\Delta E}{T}}$
13:     **end if**
14:   **until** the number of accepted iterations is below a threshold level
15:   Set $T = T - \Delta T$
16: **end while**
17: Set Best solution as x

Figure 4.3 Simulated annealing pseudocode.

temperature of the heat bath slowly trying to let atoms arrange themselves orderly. Similar, the SA algorithm starts at a heated state, moving through states that decrease the energy and some "bad" moves that are allowed if they are less than a probability that depends on the system's temperature. As the temperature drops, "bad" moves are less and less accepted, until it reaches a certain temperature that only steps that decrease the energy are permitted.

This behavior is simulated by each iteration making a random move to the state and a difference in energy is calculated as ΔE. Then, if $\Delta E \leq 0$, the move is accepted. On the contrary, if $\Delta E > 0$, the move will only be accepted using the Boltzmann distribution $p(\Delta E) = exp(-\Delta E/k_B T)$, where k_B is the Boltzmann constant and T is the temperature. In many cases, a step size α is included depending on whether the space is large or small. The pseudocode of SA is shown in Figure 4.3.

4.3 TABU SEARCH

Taboo is an implicit prohibition constructed on cultural senses that are undesirable or, perhaps, too inviolable. The Tabu search algorithm [27,28] is based on local search with the variations that permit us to move to possible worsen direction and prohibit coming back to already-searched positions using a recall (or Tabu) lists.

Tabu lists work as a type of memory that has a history of earlier searches. It is worth noticing that the memory stored is finite to a certain quantity to confine previous movements for a limited time. This is performed so it does not go back to previously visited places, while still being able to make a move in a later stage that can benefit the search and, therefore, make a larger examination.

There exist some cases in which a Tabu requires to be canceled since it can restrict an appealing movement, and hence the aspiration criteria. One aspiration criterion is to cancel a Tabu if the next movement will improve the solution and the place has not been visited previously.

The Tabu search algorithm mainly involves four steps: forbid, freeing, short-term memory, and surpass, which (1) regulates what is set in the tabu list, (2) regulates what is removed, (3) interacts on setting and removing, and (4) uses aspiration criteria if a better solution is found, respectively.

The algorithm can be described using the following steps:

- Select all near solutions.

- From the list, select the best solution that is not a Tabu.

- If it is on the list, check if it can be ignored using aspiration.

- Record the movement on the tabu list.

- Return to step one.

In Figure 4.4, there is an example of a Tabu list with a size of three that it is trying to find its minimum. The pseudocode of tabu search is shown in Figure 4.5.

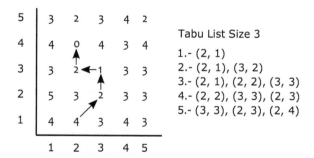

Figure 4.4 Tabu search example.

1: Set Objective Function $f(x)$, $x = [x_1, \ldots, x_d]$.
2: Select Tabu list size
3: Initialize the first solution S_0
4: Set first solution as best solution $S^* = S_0$
5: Initialize the tabu list as empty
6: **while** $t =< Max\ iterations$ **do**
7:    Select a next solutions in the neighbor.
8:    From the solutions select the one that is the minimum $\hat{S}$
9:    **if** $f(\hat{S}) > f(S^*)$ **then**
10:       set $S^* = \hat{S}$ and $f^* = \hat{f}$
11:    **end if**
12:    Record tabu for the current move
13:    Remove oldest entry of tabu list if maximum size has been reached
14: **end while**
15: Set Best solution as S^*

Figure 4.5 Tabu search algorithm pseudocode.

CHAPTER 5

Memetic Algorithms

5.1 ANT COLONY OPTIMIZATION

Ant colony optimization (ACO) [14,16] is a global optimization method that allows us to search for an optimal route between two states through possible solutions encoded as traces of pheromones.

In nature, ants find the shortest path to food by putting a pheromone trail from the nest to the source. As this trail is further traveled, the shortest route will increase the pheromone level sooner than the longer route. In the long run, the shortest route will have been traveled, thus more increasing its pheromone level to a degree that most ants will prefer this route compared to the longer route.

The algorithm can be simply described as follows:

- Initialize an ant colony.

- Initialize pheromone trails and random attraction levels.

- Repeat until a termination criterion.

    - Choose for each ant a path with a probability P.
    - Advance to the next chosen state.
    - Update the traces of pheromones of ants.
    - Update pheromone attraction levels.

- Return the best pheromone trail.

Every ant will move from node i to node j with probability,

$$p_{i,j} = \frac{(\tau_{i,j}^{\alpha})(\eta_{i,j}^{\beta})}{\sum (\tau_{i,j}^{\alpha})(\eta_{i,j}^{\beta})} \tag{5.1}$$

```
1: Set Objective Function f(x), x = [x₁,...,xₐ].
2: Initialize ant tours τ
3: Define α, β, ρ and Q
4: while t =< Max iterations do
5:     for each ant in the colony do
6:         Set all nodes as unvisited
7:         while Number of unvisited nodes > 0  do
8:             Compute probability of going to unvisited nodes using 5.1
9:             Select next node according to probability
10:        end while
11:    end for
12:    Update pheromone level 5.2 and 5.4
13: end while
14: Best solution with the shortest distance L
```

Figure 5.1 Ant colony algorithm pseudocode.

where $\tau_{i,j}$ is the amount of pheromone on edges i and j, α is the parameter to control the influence of $\tau_{i,j}$, $\eta_{i,j}$ is how desirable are the edges i and j, and β is the parameter to control the influence of $\eta_{i,j}$. Further, the amount of pheromone deposited is determined according to the equation

$$\tau_{i,j} = (1 - \rho)\tau_{i,j} + \Delta\tau_{i,j} \tag{5.2}$$

where ρ is the rate of pheromone evaporation and $\Delta\tau_{i,j}$ is the amount of pheromone deposited, which is typically given by

$$\Delta\tau_{i,j} = Q/L_k \quad \text{if ant k travels on edge i,j} \tag{5.3}$$
$$= 0 \quad \text{otherwise} \tag{5.4}$$

where Q is the pheromone deposited constant, and L_k is the cost of the kth ant's tour. Ant colony algorithm pseudocode is shown in Figure 5.1.

5.2 PARTICLE SWARM OPTIMIZATION

A particular type of algorithm that focuses on how species interact, especially how birds act to find food, is particle swarm optimization (PSO) [19,39]. For example, when a bird finds food, other birds approach to that location. In this transition, if a bird finds a better source of food, then other birds tend to move to that location. This process repeats over and over until birds tend to move to the location with the best source of food.

Resembling this movement, PSO starts with a population of particles (birds) that change their location following the location of the best

```
 1: Definition of objective function f(x)
 2: Initialize Particles
 3: Initialize Velocity
 4: Select Acceleration coefficients
 5: Calculate the best particle
 6: while t =< Max iterations do
 7:    for all particles do
 8:       Calculate fitness value
 9:       If the fitness value is better than the previous better set current value as the new particle best
10:    end for
11:    Choose the particle with the best fitness value as general best
12:    for all particles do
13:       Calculate particle velocity equation 5.5
14:       Update particle position according equation 5.6
15:    end for
16: end while
17: Set best individual as optimum
```

Figure 5.2 PSO pseudocode.

particle and their own, so far, best location. This will push the particles toward the best location (position with the most food).

In particular, the PSO starts with particles with same position and velocity. At this moment, the objective function is evaluated to define the best particle and each particle's best(current) location. Then, the iteration process starts by updating the velocities using acceleration parameters ϕ_1 and ϕ_2 (it is common to set $\phi_1 = \phi_2 = 2$). The formula to update the velocity is as follows:

$$vel = \omega vel + \phi_1 U_1 \left[pos^* - pos \right] + \phi_2 U_2 \left[pos^\star - pos \right] \qquad (5.5)$$

where pos and vel are the position and velocity of particles, U is the random number, ω is the inertia weight, and $*$ and $\star$ are the current particle best location and the global particle best location. Using this velocity, each particle's location can be expressed as:

$$pos = pos + vel \qquad (5.6)$$

Both equations will push all particles to the current best location, yet, during this movement, they may find a better location, and at that moment, all particles will tend to move to that location. The pseudocode for PSO is shown in Figure 5.2.

5.3 BAT OPTIMIZATION

Bat optimization is a metaheuristic algorithm inspired by micro-bats used for global optimization. The algorithm is based on the echolocation of micro-bats and was developed in 2010 [87,88].

Micro-bats use echolocation to navigate in the dark. The micro-bats produce a pulse in a definite angle and listen the echo that bounces back from the surrounding objects. According to how the echo returns, they can detect prey or objects, their speed, and their distance. The sound changes emission pulse rates and loudness.

The echolocation of micro-bats is summarized as follows: Each bat starts randomly with a velocity v_i at position x_i with a varying frequency or wavelength and loudness A_i. As it finds a prey, it changes frequency, loudness, and emission rate r_i. The search intensifies by walking locally, and the selection of the best bat goes on until the stopping criterion is reached. The equations to determine the frequency, velocity, and position are as follows:

$$F_i = f_{min} + (f_{max} - f_{min}) * \beta \tag{5.7}$$

$$v_i = v_i + (x_i - x_*) * F_i \tag{5.8}$$

$$x_i = x_i + v_i \tag{5.9}$$

where β is a random value from zero to 1, and x_* is the global best location. After the best solution is found, the local search starts using random walk:

$$x_{new} = x_{old} + \varepsilon A^t \tag{5.10}$$

where $\varepsilon \in [-1, 1]$ is a random variable and A^t is the average loudness of all bats at the current time step. The loudness and emission rate for each bat can be calculated as:

$$A_i = \alpha A_i \tag{5.11}$$

$$r_i = r_i^0[1 - exp(-\gamma t)] \tag{5.12}$$

where α and γ are the constants. The main advantages of the bat optimization technique are its simplicity, flexibility, and easiness to design. Furthermore, it is a very powerful tool for solving many complex problems of engineering and sciences. The pseudocode for bat optimization is shown in Figure 5.3.

5.4 GRAY WOLF OPTIMIZATION

Gray wolf optimization (GWO) is a bioinspired algorithm based on the hierarchy and hunting mechanisms of gray wolves. It was developed by Mirjalili in 2010 [46] and is based on the fact that wolves are considered apex predators; that is, they live and hunt in packs. Also, in their way,

```
1: Set Objective Function f(x), x = [x₁,...,xₐ].
2: Initialize bat population position and velocity xᵢ(1,...,n) and vᵢ.
3: Define pulse frequency fᵢ Initialize pulse rates and loudness rᵢ and Aᵢ
4: while t =< Max iterations do
5:    Adjust the frequencies 5.7
6:    Update velocities and locations 5.8 and 5.9
7:    if rand > rᵢ then
8:       Select solution among best solutions
9:       Make a local solution around the selected best solution
10:   end if
11:   Find a new solution by flying randomly
12:   if rand < Aᵢ and f(xᵢ) < f∗ then
13:      Accept new solutions
14:      Reduce Aᵢ and Increase rᵢ using 5.11 and 5.12
15:   end if
16:   Rank bats and select current best x∗
17: end while
```

Figure 5.3   Bat algorithm pseudocode.

they conduct themselves: An alpha wolf is the leader and decision-maker; beta, delta, and omega wolves help the alpha wolf in the decision-making process; and the rest of the wolves are the followers.

Furthermore, hunting process of wolves is divided into three stages: searching for prey (tracking and approaching), encircling and harassing the prey, and finally attacking prey [21].

Analogously, GWO considers the alpha α, the beta β, and delta δ wolves the main, the second, and the third best solutions, respectively, and the rest of the solutions are the omega ω wolves. Also, similar to the behavior of wolves, the α, β, and δ wolves are involved in the hunting process.

The first process is tracking and approaching the prey, represented as:

$$D = |C \cdot X_p - A \cdot X| \tag{5.13}$$

$$X = X_p - A \cdot D \tag{5.14}$$

where X_p and X are the position of prey and the position of gray wolf, respectively. Moreover, A and C are the coefficient vectors, represented as

$$A = 2ar_1 - a \tag{5.15}$$

$$C = 2r_2 \tag{5.16}$$

where r_1 and r_2 are the random values between $[0, 1]$, and a is the vanishing value from 2 to 0.

The second gray wolf hunting process is encircling and harassing the prey. In this process, the wolves locate and surround the prey. Hence, the algorithm uses the positions of the alpha α, beta β, and delta δ wolves, and the rest of the wolves (omegas ω) are forced to update their position depending on the information. This update is done as follows:

$$D_\alpha = |C_1 \cdot X_\alpha - A \cdot X|$$
$$D_\beta = |C_2 \cdot X_\beta - A \cdot X| \tag{5.17}$$
$$D_\delta = |C_3 \cdot X\delta - A \cdot X|$$

$$X_1 = X_\alpha - A_1 \cdot D_\alpha$$
$$X_2 = X_\beta - A_2 \cdot D_\beta \tag{5.18}$$
$$X_3 = X_\delta - A_3 \cdot D_\delta$$

$$X = \frac{X_1 + X_2 + X_3}{3} \tag{5.19}$$

The final process is exploitation and exploration; accordingly, wolves attack their prey and start searching for new and better prey. For this process, the algorithm uses the value A that goes from $[-2a, 2a]$ (keeping in mind that a goes from 2 to 0); when $|A| < 1$, the wolves attack toward the prey (exploitation). On the other hand, when $|A| > 1$, the wolves diverge from the prey to find a more suitable one (exploration).

The algorithm for the hunt is shown in Figure 5.4.

1: Definition of objective function $f(x)$
2: Set Population size n
3: Set A and C coefficients
4: Number of iterations t
5: Initialization of gray wolf population X_i
6: Calculate the fitness to each wolf f_i
7: Choose the best solutions
8: Set X_α, X_β and X_δ as the first, second and third best solutions respectively.
9: **while** $t =< Max\ iterations$ **do**
10:　　For each wolf update the position
11:　　Update a, A, C
12:　　Calculate the fitness to all population f_i
13:　　Update $X_\alpha, X_\beta, X_\delta$
14: **end while**
15: Set solution as current global best X_α

Figure 5.4　GWO algorithm pseudocode.

CHAPTER 6

Geological Optimization

M ETAHEURISTIC ALGORITHMS ARE CLASSIFIED into four principal categories: (1) evolutionary based, (2) physical based, (3) swarm based, and (4) human behavior based [52]. First, evolutionary-based algorithms imitate the natural biological evolution and/or social behavior of species [20]. Some examples of this category are genetic algorithm (GA) [32], evolutionary programming (EP) [23], evolutionary strategy (ES) [68], and differential evolution (DE) [76].

Second, physical-based algorithms have some characteristics of the laws and forces of physics such as gravitational, electromagnetic, and inertia. Those rules govern the communication or movement behaviors between the search agents [13]. Some of the most popular physical-based algorithms are simulated annealing (SA) [40], gravitational search algorithm (GSA) [67], magnetic optimization algorithm (MOA) [78], charged system search (CSS) [38], central force optimization (CFO), black hole (BH) [30], galaxy-based search algorithm (GbSA) [74], artificial chemical reaction algorithm (ACROA) [1], and the small-world optimization algorithm (SWOA) [18].

Third, swarm-based algorithms use the collective intelligence from organisms in a community sharing local information with other individuals, such as ants and bees [84]. The most known methods include particle swarm optimization (PSO) [19], monkey search algorithm (MSA) [50], bat algorithm (BA) [87], cuckoo search (CS) [24], wolf pack search algorithm [83], firefly algorithm (FA) [85], artificial bee colony (ABC) [35], and the ant colony optimization (ACO) [15].

Finally, the last category is based on the human behavior in different situations or activities, such as teaching-learning-based optimization (TLBO) [66], neighborhood searching (Tabu search) [27], sociopolitically

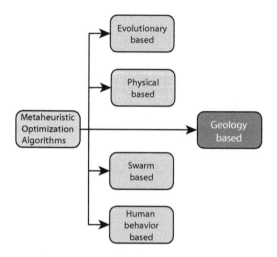

Figure 6.1 Classification of metaheuristic optimization algorithms.

motivated strategies (imperialist competitive algorithm (ICA)) [77], musicians harmony calibration for melodies (harmony search (HS)) [25], and mimicking team sports competitions (league championship algorithm (LCA)) [37].

Figure 6.1 displays a broad classification of metaheuristic optimization algorithms, highlighting the fact that the main objective of this chapter is to present a new research line in metaheuristic optimization algorithms, enabled by the novel earthquake algorithm (EA), which is the first geo-inspired algorithm.

As explained in [63], Earth (Geo) and environment phenomena can be source of inspiration for the development of a variety of applications; in this particular case, they serve as inspiration for a metaheuristic optimization algorithm, taking as backbone the behavior of S and P seismic waves.

6.1 EARTHQUAKE ALGORITHM

6.1.1 Background

Earthquakes are one of the most catastrophic and destructive natural disasters, causing thousands of deaths and economic losses [36]. In seismic zones, elastic energy accumulates around regions before friction is overcome, followed by unexpected shear movements giving rise to an earthquake; at that point, the surrounding strained energy decreases and the seismic waves are generated, which finally shakes the crust of the earth [41].

The final magnitude of an earthquake is defined using the P-wave, which provides enough information after the earthquake begins [59]. Therefore, the velocity of the P-wave can be used to make a good exploration by delivering information from the search space to find the optimum solution.

On the other hand, [7] explains that seismic waves have elastic properties, which may change according to the medium (earth materials) they are moving on. In other words, the degree of elasticity dictates how the waves are transmitted.

Also, during an earthquake, earth materials are subjected to different types of stresses such as compression, tension, and/or shearing forces. In earth layers, some materials are ductile at very slow strain rates, and movements are on the order of [mm] or [cm] per year. Meanwhile, subsoil materials react elastically to fast and small deformations produced by earthquakes; therefore, when high-amplitude seismic deformations with long periods occur, free-oscillation modes show up and inelastic responses of the seismic waves must be considered.

According to Ref. [8], elastic rates can be defined by Hooke's law for subsoil materials. Above their elastic limits, the materials may suffer brittle fracturing (e.g., structural earthquake failures) or ductility [2]. Nevertheless, Ref. [7] explains that depending on the types of deformation, the elastic material endures stress and can be quantified by the elastic moduli:

- The *bulk modulus*, κ, is described as the ratio of the hydrostatic pressure change to the resulting relative volume (V) change (i.e., $\kappa = \Delta P/(\Delta V/V)$).

- The *Lamé parameter*, μ, *(shear modulus)* is the resistance of the material to shearing, namely, to change the shape and not volume (i.e., $\mu = \tau_{xy}/2e_{xy}$ or $\mu = (\Delta F/A)/(\Delta L/L)$).

- The *Young modulus*, *E*, is defined by the response of a cylinder to length that is pulled or pressed on both ends. Its value is given by $E = (F/A)/(\Delta L/L)$.

- *Poisson's ratio*, σ, is the ratio between a cylinder's lateral contraction as it's being pulled on its ends and its relative longitudinal extension (i.e., $\sigma = (\Delta W/W)/(\Delta L/L)$).

- *Lamé parameter,* λ, does not have a physical explanation, but it can be defined in terms of the elastic moduli already mentioned $(\lambda = \sigma E/((1+\sigma)(1-2\sigma)))$.

- The *density,* ρ, of the earth material: Denser rocks have faster wave propagation because rigidity increases with density.

6.1.2 P- and S-Wave Velocities

In a real earthquake, the rocks break and generate waves through the subsoil and to the surface of the earth, as shown in Figure 6.2. However, as explained before, there are two types of earthquake waves, namely, P and S, that are generated. The P-wave is fast and depends on earth material compressibility. These waves are transmitted by compression and tension of the medium with volume changes. The S-waves are slower than the P-waves, and their transmission depends on rock elasticity; the epicenters are moved up and down perpendicular to the wave propagation direction.

Therefore, in order to gain a complete understanding on how the waves appear in earthquakes, Figure 6.3 shows the ground acceleration data recorded from an earthquake in Chiapas, Mexico, on 15 September 2010, where the epicenter originated in the coordinates 15.59° N, 93.52° W at 95 km deep (according to Ref. [80]). Also in Figure 6.3, P- and S-waves are highlighted, and hence, it can be analyzed how the P-waves appear first, due to its higher velocity; meanwhile, S-waves appear second, identified in the data collected by the accelerometers.

On the one hand, the movement of the P-waves through a medium is made by compression and dilation (as shown in Figure 6.4). These types of earthquake waves can be propagated through any medium (solid, liquid, and gas).

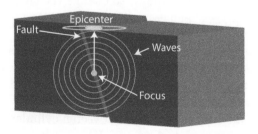

Figure 6.2 Origin of an earthquake.

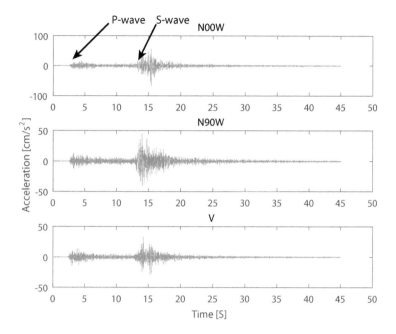

Figure 6.3 Ground acceleration (three-axis) recording during an earthquake.

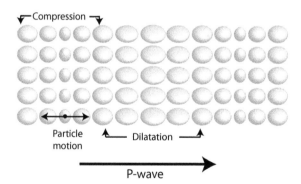

Figure 6.4 *P*-wave traveling through an earth material by compression and dilation.

On the other hand, the propagation of the *S*-waves can only be made through solids, where it induces shearing deformation in the propagating medium (perpendicular movement from the propagation direction). The behavior of the *S*-waves is shown in Figure 6.5.

To understand the relation of the propagation waves with different geologic formations, Table 6.1 shows the main density values, Poisson's

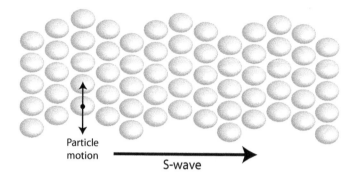

Figure 6.5 *S*-wave traveling through an earth material by shearing deformation.

TABLE 6.1 Principal Materials or Geologic Formation Properties [7].

Material or Geologic Formation	Density (Kg/m^3)	Poisson's Ratio	v_p/v_s
Sandstone	2500	0.21	1.65
Salt	2200	0.17	1.59
Limestone	2700	0.19	1.62
Granite	2610	0.25	1.73 (1.65-1.91)
Basalt	2940	0.28	1.8 (1.76-1.82)
Peridotite, Dunite, Pyroxenite	3300	0.29	1.8 (1.76-1.91)

ratio, and the ratio of seismic wave velocities from some earth materials (as described in Ref. [7]).

6.1.3 Earthquake Optimization Algorithm

As explained above, the (*EA*) is a geo-inspired algorithm, based on the behavior of the *P*- and *S*-waves existing in earthquakes. On September 19, 2017, Mexico City suffered the impact of an enormous earthquake that showed up all the energy that is released in a short period. This significant event inspired to Dr. Pedro Ponce and Dr. Arturo Molina to work on a new structure of metaheuristic optimization based on the description of an earthquake. Thus, they coined the term earthquake optimization [63]; meanwhile, the modified version of the algorithm was presented and implemented in [45], showing its implementation for the optimization of a proportional integrative derivative (PID) controller for electric machines.

Therefore, summarizing the key concepts from the natural phenomenon for the deduction of the earthquake optimization algorithm, the P-waves are faster and the S-waves are finer. Accordingly, equations (6.1) and (6.2) describe the estimation of P- and S-wave velocities.

$$v_p = \sqrt{\frac{\lambda + 2\mu}{\rho}} \tag{6.1}$$

$$v_s = \sqrt{\frac{\mu}{\rho}}, \tag{6.2}$$

where v_p and v_s are the P- and S-wave velocities, respectively, λ and μ are the Lamé parameters, and ρ is the density of the earth material where it is propagating. Then, as explained in [41] and [45], equations (6.1) and (6.2) show the mathematical expressions used for the P- and S-wave velocities, which are the backbone of the algorithm.

Therefore, according to [7], the Lamé parameters can be the same under some circumstances; hence, [45] explains that for the EA, it is taken that $\lambda = \mu$. Nevertheless, [47] explains that in order to find the best Lamé parameters to be used, several tests are performed with different Lamé values, finding that the only real constant that worked is 1.5, which is explained by:

$$\lambda = \mu = 1.5 \, GPa \tag{6.3}$$

To validate the proposed constant for the Lamé parameters, [7] explains that most of the rocks have a *poisson ratio* between 0.2 and 0.3, leaving a mean optimal value of 0.25 for the ratio. Then, the relation between Poisson's ratio and the Lamé parameters, according to Ref. [8], is shown by (6.4).

$$\sigma = \frac{\lambda}{2(\lambda + \mu)}, \tag{6.4}$$

where σ is Poisson's ratio. Thus, substituting (6.3) in (6.4) results in $\sigma = 0.25$, which validates a correct selection for (6.3).

Nevertheless, the density of the solids (ρ) is used as a random value, which gives the heuristic behavior to the algorithm; so as not to leave the geo-inspiration of the algorithm, the selected range for the density parameter is taken between 2200 and 3300 Kg/m^3, taken from the geological properties described in Table 6.1.

After defining the main parameters of the algorithm, the use of v_s or v_p is not yet clarified. Thus, in order to determine when to use one or the other, it is essential for the algorithm to define an operation range

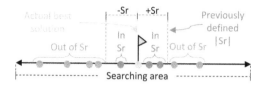

Figure 6.6 Procedure to find particles within and beyond the defined Sr.

for the v_s to be used, which, as explained in [8,36,41] and [45], is defined as the *S-range* or *Sr*. The *Sr* as shown in Figure 6.6 is defined always around the best solution.

Therefore, the actual behavior of the algorithm "orbits" the global best solution; however, the *Sr* should be assigned (as explained in Ref. [45]) with the previous knowledge of the problem requirements, but it is recommended to implement the range in function of the percentage of error between the obtained solutions and the expected ones. In applications presented in [36] and [45], the Sr is recommended to be 2% from the best solution.

Notwithstanding, it is important to highlight the fact that since both speeds are calculated by a square root, the final result is a positive number. The architecture of the algorithm contemplates the random selection of a positive or a negative velocity value, regardless of whether it is v_s or v_p.

In addition to the algorithm, the update of the current position of the epicenters is given by equation (6.5):

$$X_i^t = X_i^{t-1} + V_i, \tag{6.5}$$

where X_i^t and X_i^{t-1} are the current and previous positions, respectively, and V_i is the current velocity (v_s or v_p).

And finally, as explained in [45], a random selection for an exponential distribution reduces the probability of visiting points already visited for the epicenters, where the implemented expression of the distribution is taken from [10] (according to its relation with the poisson distribution). The random value is generated in a range of ± the maximum value of v_p/v_s, that is, ± 1.91 (taken from Table 6.1), resulting in the relation described by (6.6).

$$X_i^t = X_{best} + Exp_\mu(s), \tag{6.6}$$

where X_{best} is the global best solution and $Exp_\mu(s)$ is the random value generated with the exponential distribution from the value of μ.

Figure 6.7 shows a synthesized architecture of the EA, where the flowchart makes it clear to visualize where the decision of whether to use

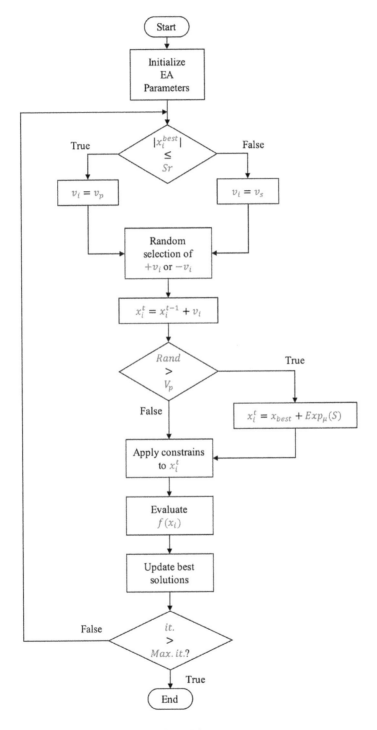

Figure 6.7   General architecture for *EA*.

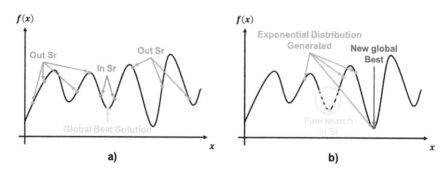

Figure 6.8 (a and b) Example of the EA behavior with 10 epicenters.

v_p or v_s is made, where the additional heuristic freedom degree with the exponential distribution is added, and where the solutions are evaluated and ranked. A more detailed flowchart was found in Ref. [45].

Regarding the final generation of some epicenters with the random exponential distribution, the additional freedom degree given by those heuristic "moves" helped the algorithm to send a couple of epicenters to a fast search, when it seemed that the group of epicenters started to look for a convergence in a local minimum. That additional generation of epicenters improved the escaping capabilities of the algorithm from local solutions.

An example to clarify how the EA works is shown in Figure 6.8, where the generalities of the behavior are analyzed. Figure 6.8a shows an example of how the epicenters can be randomly placed on a function ($f(x)$). As explained in [45], after ranking the solutions to find the first global best, the epicenters that are in or out of the Sr can be known, and hence, those that are going to use the v_s (epicenter in of Sr) or v_p (epicenter out of Sr), in order to update their positions.

Meanwhile, Figure 6.8b shows the searching behavior where it can be seen that after some iterations, the epicenters start to converge with a fine search around the current global best (as also explained in Ref. [45]). From the same figure, it can be observed that four epicenters took distant routes from the epicenters set, due to the exponential distribution random generation, allowing the algorithm to escape the local minimum where the other epicenters were getting trapped.

CHAPTER 7

Optimization MATLAB App and LabVIEW Toolkit

7.1 MATLAB APP

The MATLAB® app is classified into two main apps: the mainAPP.mlapp and mainAPPTSP.mlapp, which are also referred to as primary user interface (PUI) and secondary user interface (SUI), respectively. These two apps can be run by double-clicking on each of these files or by opening the main.m or mainTSP.m, respectively, and pressing F5 (run). The four files are shown in Figure 7.1.

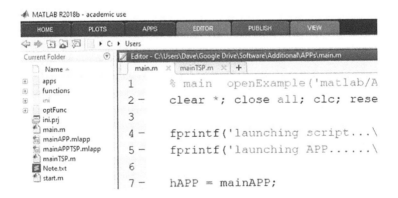

Figure 7.1 Opening the main.m file running the program.

Figure 7.2 (a and b) Main and secondary APP.

After either app is opened, a main user interface will appear (see Figure 7.2a for PUI and Figure 7.2b for SUI).

The first part to observe in both user interfaces is the drop-down menu where the optimization algorithm can be selected. The PUI contains the algorithms such as Nelder–Mead (NM), genetic algorithm (GA), simulated annealing, particle swarm optimization (PSO), bat algorithm (BA), gray wolf optimization (GWO), and earthquake algorithm (EA). On the other hand, the SUI contains the algorithms such as ant colony optimization (ACO) and Tabu search optimization. These algorithms are also divided into two parts: the algorithms that are normally continuous and the algorithms that are normally discrete.

7.1.1 Primary User Interface

Once the method has been selected, the optimization function can be inserted into the function "edit field text". In this interface, it can be typed either a new function, for instance, $f(x, y) = x^2 + 2xy + y^2 + 10$, or as $x(1)\hat{\ }2 + 2x(1)x(2) + x(2)\hat{\ }2 + 10$. Alternatively, the name of a file containing a function can be inserted that has other variables to study. Some examples can be found under the folder ./$optFunc$ (for further information on these functions, see Section 8.1).

Additionally, within the PUI, three more variables can be defined: the total number of variables of the optimization function, the maximum number of epochs, and the optimization function limits. It is worth noticing that the limits will change according to the number of variables selected and it will add as many limits as variables are selected. Further, each limit should have the format $[min; max]$. The user interface will not allow to continue if these format is not follow as presented or it will invert the limit if the limits are inserted backward.

After all the variables of PUI have been selected, the submit button must be clicked to move to each algorithm user interface.

7.1.1.1 Algorithm User Interfaces

Before elaborating on each algorithm user interface, it is important to notice that all of them have some similarities. First, at the top of each user interface, the name of the algorithm is displayed; on the left, the objective function is rewritten according to what was written in the main user interface. Also, there is a result box at the bottom where the location of optimum value and its value are going to be displayed. Lastly, there are two graphs that will display the optimum value for each iteration and how the optimization variables change.

a. The NM user interface (Figure 7.3a) requires selecting α, γ, ρ, and σ, and their values can be modified either by inserting a new number within the spinners or by clicking on the arrows.

b. GA (Figure 7.3b) requires selecting the cross and mutation probability using a spinner and how many individuals each generation would have. These values start with a probability of 50%.

c. The simulated annealing requires selecting the number of iterations (or number of steps) each temperature will take for a single temperature, the initial and final temperature, and the step size α (Figure 7.3c).

d. PSO user interface requires selecting how many particles will seek for the best solution, the inertia weight ω, and the acceleration parameters ϕ_1 and ϕ_2 (Figure 7.3d). Also, it is recommended to select the velocity limits so the particles do not exceed a certain threshold.

e. The BA user interface requires selecting the number of bats and the constants α and γ (Figure 7.3e). Further, it requires selecting the frequency limits and the velocity limits of the bats. Note that the initial emission rate Ro and loudness A are initialized randomly for each bat.

f. For the GWO, the only variable that requires to be selected is the total number of wolves (Figure 7.3f).

g. Similar to GWO, the EA only requires selecting the number of epicenters as initial parameter (Figure 7.3g).

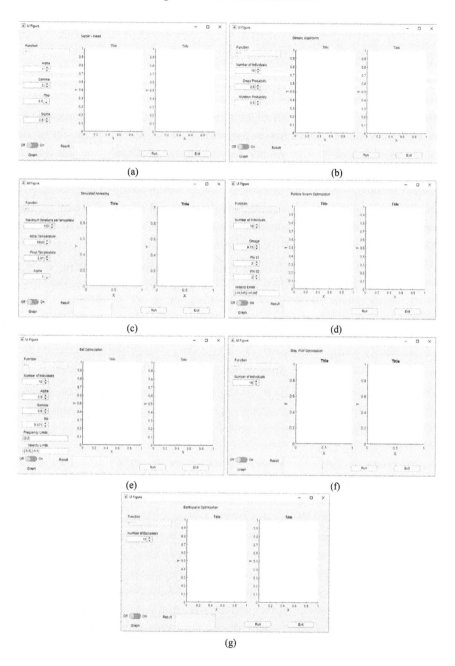

Figure 7.3 Optimization algorithms: (a) NM, (b) GA, (c) simulated annealing, (d) PSO, (e) BA, (f) GWO, and (g) EA.

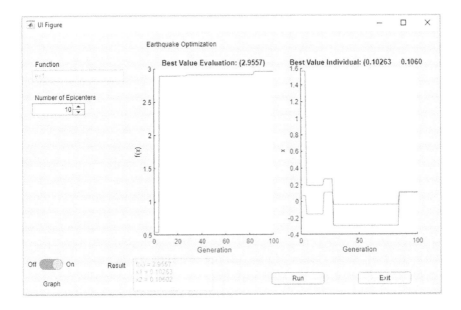

Figure 7.4 Example on how to use one of the applications (earthquake).

An example on how to use one of the applications is shown in Figure 7.4, where 10 epicenters are selected and the evolution of the iterations is from 0.5 to ≈ 3 with final location ≈ 0.102 and ≈ 0.102.

7.1.2 Secondary User Interface

For the case of the SUI, it requires inserting the points. For example, to insert the coordinates $x_1 = [5, 5]$, $x_2 = [4, 3]$, and $x_3 = [2, 1]$, it should be typed as $[5, 5; 4, 3; 2, 1]$. Alternatively, the name of a file containing those points can be inserted (e.g., $tsp01.m$). Also, the cost function should be inserted in the second box. As the initial cost function, it uses the norm (distance.m) but any other file can be used just by inserting its name in the box. Some examples can be found under the folder $./optFunc$ (for further information on these functions, see Section 8.1).

7.1.2.1 Algorithm User Interfaces

Similar to PUI, the algorithms in SUI share some features. First, at the top of each user interface, the name of the algorithm is displayed; also on the left, the discrete points or file name is rewritten according to what was written in the main user interface. Further, there is a result

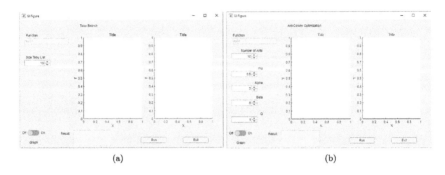

Figure 7.5 (a) Tabu search and (b) ant colony optimization.

box at the bottom where the discrete points are going to be rewritten with the final order found. Lastly, there are two graphs that will display each iteration optimum value and how the order of the discrete points changes.

a. The Tabu search user interface requires selecting the size of the tabu list using the spinners or by typing the value (Figure 7.5a).

b. The ACO user interface requires selecting the number of ants, α and β as parameters to control for the pheromone influence and edge desirability, ρ as the evaporation rate, and Q as the pheromone deposited constant (Figure 7.5b).

An example on how to use one of these applications is shown in Figure 7.6, where 18 cities were used as input (see 8.1 Project 08 - tsp01) and the evolution of the iterations is from ≈ 775 to ≈ 390.

7.1.3 Individual Functions

Alternatively, each of the optimization algorithm can be run without the app. To do this, it is important to insert the hyperparameters required. An example on how to use them can be found on $./functions/methods$ on $test.m$ and $testTSP.m$. The algorithms for the PUI and their inputs are described as follows:

a. The NM function [maxPop, maxVal] = NMO(Nd, alpha, gamma, rho, sigma, fun, posLimit, maxEpochs, hB, hP, pri), with Nd as the number of dimensions of the optimization function, alpha as the reflection coefficient, gamma as the expansion coefficient, rho as the contraction coefficient, sigma as the shrinking coefficient,

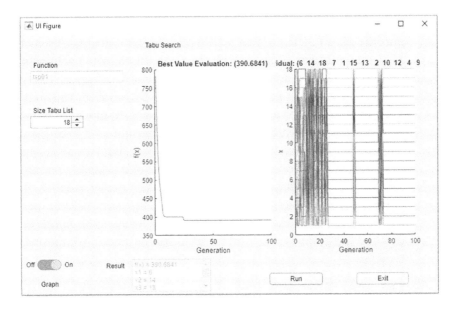

Figure 7.6 TSP example with 18 cities. The algorithm (Tabu search) was run over 100 iterations.

fun as the optimization function, posLimit as the research space limits, maxEpochs as the maximum number of epochs to run the algorithm, hB as the handle axis to plot current optimum, hP as the handle axis to plot current best position, pri as the authorized printing on both of the previous axis, maxPop as the best position found, and maxVal as the best optimum value found.

b. GA [maxPop, maxVal] = GAO(N, Nd, Pc, Pm, fun, posLimit, maxEpochs, hB, hP, pri), with N as the number of individuals, Nd as the number of dimensions of the optimization function, Pc as the percentage of crossing, Pm as the percentage of mutation, fun as the optimization function, posLimit as the research space limits, maxEpochs as the maximum number of epochs to run the algorithm, hB as the handle axis to plot current optimum, hP as the handle axis to plot current best position, pri as the authorized printing on both of the previous axis, maxPop as the best position found, and maxVal as the best optimum value found.

c. The simulated annealing function [maxPop, maxVal] = SAO (maxIter, Nd, Tini, Tfin, alpha, fun, posLimit, maxEpochs, hB, hP, pri), with maxIter as the maximum number of iterations per temperature(epoch), Nd as the number of dimensions of the optimization function, Tini as the initial temperature, Tfin as the final temperature, alpha as the step size, fun as the optimization function, maxEpochs as the maximum number of epochs to run the algorithm, posLimit as the research space limits, velLimit as the research space velocity limits, Fm as the frequency limits, A as the average loudness of all bats (random number from zero to one), r as the initial emission rate for each bat (random number from zero to one), alpha as the constant alpha, gamma as the constant gamma, hB as the handle axis to plot current optimum, hP as the handle axis to plot current best position, pri as the authorized printing on both of the previous axis, maxPop as the best position found, and maxVal as the best optimum value found.

d. The PSO function [maxPop, maxVal] = PSO(N, Nd, fun, maxEpochs, posLimit, velLimit, omega, phi, hB, hP, pri), with N as the number of particles, Nd as the number of dimensions of the optimization function, fun as the optimization function, maxEpochs as the maximum number of epochs to run the algorithm, posLimit as the research space limits, velLimit as the research space velocity limits , omega as the inertia weight, phi as the acceleration parameters, hB as the handle axis to plot current optimum, hP as the handle axis to plot current best position, pri as the authorized printing on both of the previous axis, maxPop as the best position found, and maxVal as the best optimum value found.

e. The BA function [maxPop, maxVal] = BAT(N, Nd, fun, maxEpochs, posLimit, velLimit, Fm, A, r, alpha, gamma, hB, hP, pri), with N as the number of bats, Nd as the number of dimensions of the optimization function, fun as the optimization function, maxEpochs as the maximum number of epochs to run the algorithm, posLimit as the research space limits, velLimit as the research space velocity limits , Fm as the frequency limits, A as the average loudness of all bats (random number from zero to one), r as the initial emission rate for each bat (random number from zero to one), alpha as the constant alpha, gamma as the constant gamma, hB as the handle axis to plot current optimum, hP as the handle axis to plot current best position, pri as the authorized printing on both of the previous axis, maxPop as the best position found, and maxVal as the best optimum value found.

f. The GWO function [maxPop, maxVal] = GWO(N, Nd, fun, maxEpochs, posLimit, hB, hP, pri), with N as the number of individuals, Nd as the number of dimensions of the optimization function, fun as the optimization function, posLimit as the research space limits, maxEpochs as the maximum number of epochs to run the algorithm, hB as the handle axis to plot current optimum, hP as the handle axis to plot current best position, pri as the authorized printing on both of the previous axis, maxPop as the best position found, and maxVal as the best optimum value found.

g. The EA function EQO(ne, Nd, fun, maxEpochs, posLimit, buscar, hB, hP, pri), with ne as the number of epicenter, Nd as the number of dimensions of the optimization function, fun as the optimization function, maxEpochs as the maximum number of epochs to run the algorithm, posLimit as the research space limits, buscar that allows us to search for a maximum or a minimum (zero or one), hB as the handle axis to plot current optimum, hP as the handle axis to plot current best position, pri as the authorized printing on both of the previous axis, maxPop as the best position found, and maxVal as the best optimum value found.

The algorithms for the SUI and their inputs are described as follows:

a. The Tabu search function [maxPop, maxVal] = TS(cities, cFun, maxEpochs, TSize, hB, hP, pri), with cities as the position of the cities to visit, cFun as the cost function, maxEpochs as the maximum number of epochs to run the algorithm, TSize as the size of the tabu (maximum tabu size is the number of cities), hB as the handle axis to plot current optimum, hP as the handle axis to plot current best position, pri as the authorized printing on both of the previous axis, maxPop as the best position found, and maxVal as the best optimum value found.

b. The ACO function [maxPop, maxVal] = ACO(nAnts, pos, cFun, maxEpochs, rho, alpha, beta, Q, hB, hP, pri), with nAnts as the number of ants , pos as the position of the cities to visit, cFun as the cost function, maxEpochs as the maximum number of epochs to run the algorithm, rho as the evaporation rate, alpha as the pheromone influence, beta as the pheromone edge desirability, Q as the pheromone deposit constant, hB as the handle axis to plot current optimum, hP as the handle axis to plot current best position, pri as the authorized printing on both of the previous axis, maxPop as the best position found, and maxVal as the best optimum value found.

7.1.4   MATLAB Simulink

7.1.4.1   MPPT Simulink Models

The Simulink® model, as shown in Figure 8.15, shows how the actual simulation was performed using the *Specialized Power Systems* components from the *Simscape* library. The results show how the metaheuristic-based MPPTs (maximum power point trackings) improve the energy harvesting of the selected photovoltaic panel (PV) arrays under variable irradiations (Figure 7.7).

Nevertheless, for optimal operation of PV modules, the evolution of MPPT algorithms seeks to find faster and more accurate responses, in order to maximize solar energy transformed into electricity. The classic Perturb and Observe (P&O) is widely used because of its simplicity although it doesn't usually get the fastest response and usually has more variations when approaching the MPPT.

Figure 7.8 shows the obtained power through time, comparing the performance of the algorithms. In both metaheuristic cases, an improvement in establishment time and accuracy of the MPPT was demonstrated, showing the relevance of the optimization algorithms in power electronic applications.

7.2   LABVIEW APP - FRONT PANELS

The LabVIEW Apps were developed to make an interface between the optimization algorithms and the user. For each algorithm, a LabVIEW App was developed with a similar interface. These Apps are composed of input and output elements. The input elements are (1) benchmark function selector, (2) formula input (string), (3) variables used, (4) problem definition, (5) optimization algorithm parameters, and (6) start optimization. The output elements are (7) global best position, (8) best cost, (9) density graph of the objective function, (10) population positions, and (11) best cost-by-iteration graph (Figure 7.9).

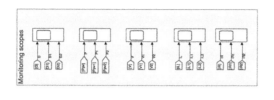

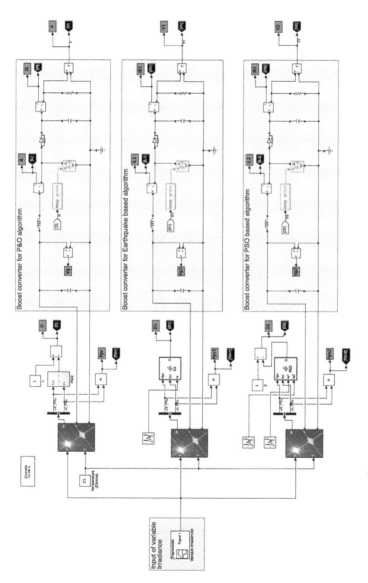

Figure 7.7 Simulink® model configuration for MPPT algorithms simulation.

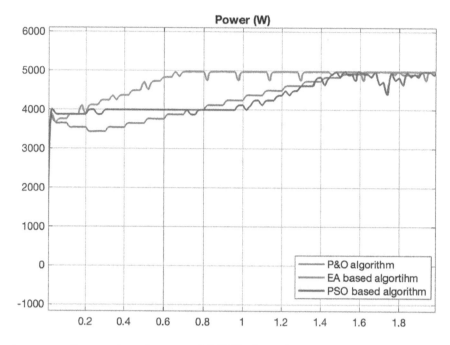

Figure 7.8 Comparison between MPPT algorithms.

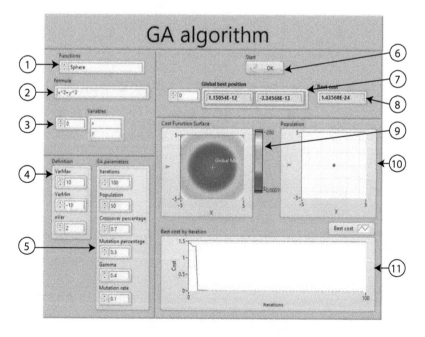

Figure 7.9 LabVIEW APP (front panel).

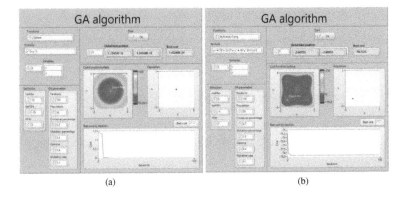

(a)   (b)

Figure 7.10 GA algorithm with (a) sphere and (b) Styblinski–Tang (front panel).

7.2.1 GA Application in LabVIEW

Figure 7.10a and b show the final results of the optimization process for GA algorithm with different benchmark functions (sphere and Styblinski–Tang).

7.2.2 PSO Algorithm Application in LabVIEW

Figure 7.11a and b depict the final results of the optimization process for PSO algorithm with different benchmark functions (sphere and surface).

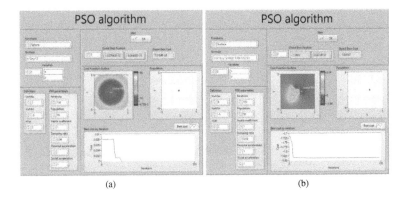

(a)   (b)

Figure 7.11 PSO algorithm with (a) sphere and (b) surface functions (front panel).

7.2.3 BA Application in LabVIEW

Figure 7.12a and b present the final results of the optimization process for BA with different benchmark functions (sphere and surface).

7.2.4 ACO Algorithm Application in LabVIEW

Figure 7.13a and b depict the final results of the optimization process for ACO algorithm with different benchmark functions (Styblinski–Tang and Keane).

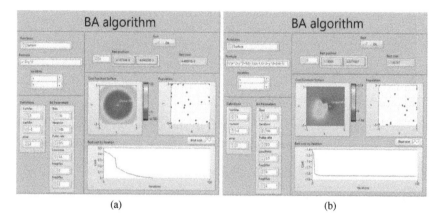

Figure 7.12 BA with (a) sphere and (b) surface functions (front panel).

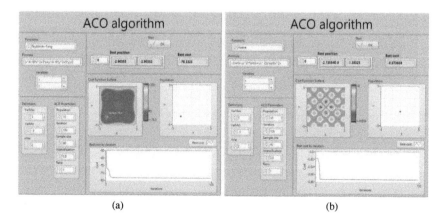

Figure 7.13 ACO algorithm with (a) Styblinski–Tang and (b) Keane functions (front panel).

7.2.5 GWO Algorithm Application in LabVIEW

Figure 7.14a and b depict the final results of the optimization process for GWO algorithm with different benchmark functions (Styblinski–Tang and surface).

7.2.6 EA Application in LabVIEW

Figure 7.15a and b present the final results of the optimization process for EA with different benchmark functions (Styblinski-Tang and surface).

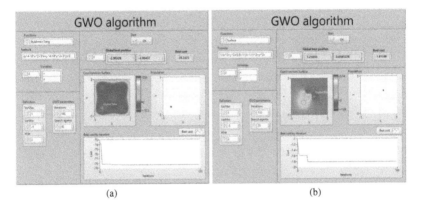

(a)   (b)

Figure 7.14 ACO algorithm with (a) Styblinski–Tang and (b) surface functions (front panel).

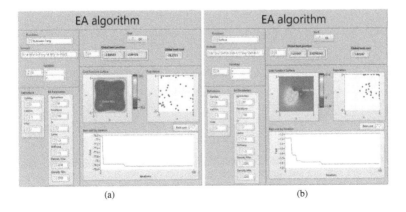

(a)   (b)

Figure 7.15 EA with (a) Styblinski–Tang and (b) surface functions (front panel).

7.2.7 NM Algorithm Application in LabVIEW

Figure 7.16a and b show the final results of the optimization process for NM algorithm with different benchmark functions (Styblinski–Tang and Keane).

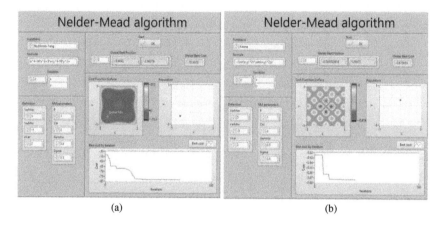

(a)   (b)

Figure 7.16 NM algorithm with (a) Styblinski–Tang and (b) Keane functions (front panel).

CHAPTER 8

Equations and Ongoing Projects

8.1 EQUATIONS

The projects can be found under the folder ./optFunc. Each of these projects can be loaded within the app just by typing its name in the function edit field text.

8.1.1 Equation 01

Project/problem description. The polynomial function $f = -x_1^2 + -3 * x_2^2 + 3$ has a maximum at $(x_1, x_2) = (0, 0)$ with value 3. This maximum is located within the intervals $[-3, 3]$ and $[-3, 3]$, which is shown in Figure 8.1.

8.1.2 Equation 02

Project/problem description. Consider the polynomial function

$$f = 3 * (1 - x_1)^2 * e^{-x_1^2 - (x_2+1)^2} - 10 * (x_1/5 - x_1^3 - x_2^5) * e^{-x_1^2 - x_2^2}$$
$$- 1/3 * e^{-(x_1+1)^2 - x_2^2}$$

This function has one global maximum $f(x_1, x_2) = f(0, 1.6) = 8.1$ and two local maximums at $f(1.4, 0) = 3.495$ and $f(-0.5, -0.6) = 3.74$. This function also has one global minimum $f(0.2, -1.6) = -6.531$ and two local minimums $f(-1.3, 0.2) = -3.032$ and $f(0.3, 0.3) = -0.0636$. These extrema are located within the intervals $[-3, 3]$ and $[-3, 3]$. These points are illustrated in Figure 8.2.

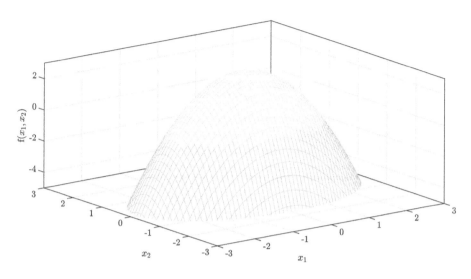

Figure 8.1   Project 01.

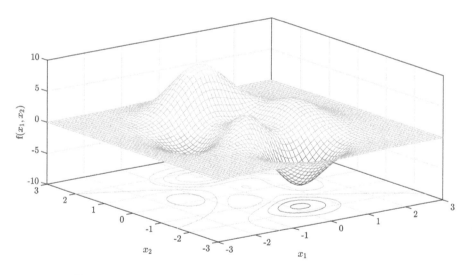

Figure 8.2   Project 02.

8.1.3   Equation 03

Project/problem description. The function $f = -(x_1^2 - 4 * x_1 + x_2^2 - x_2 - x_1 * x_2)$ has a unique maximum point at $f(3, 2) = 7$, located within the intervals $[-5, 5]$ and $[-5, 5]$ (see Figure 8.3).

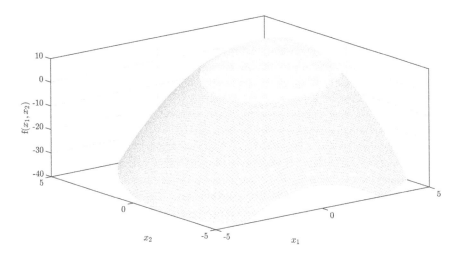

Figure 8.3 Project 03.

8.1.4 Equation 04

Project/problem description. The function $f = cos(x_1) * cos(x_2) * e^{-(x_1-\pi)^2-(x_2-\pi)^2}$ is an oscillating function with several local maximums and minimums, yet it only has one global maximum $f(\pi, \pi) = 1$ located in the intervals $[0, 5]$ and $[0, 5]$, (see Figure 8.4). Note that the oscillations within these functions are not easily observed and also that oscillations have a flat neighborhood that in some cases slows the convergence.

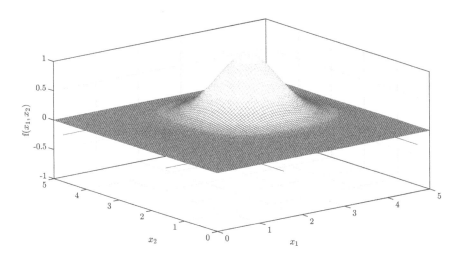

Figure 8.4 Project 04.

8.1.5 Equation 05

Project/problem description. A company manufactures two machines, namely, A and B. Using available resources, either 28 A or 14 B can be manufactured daily. The sales department can sell up to 14 A machines or 24 B machines. The shipping facility can handle no more than 16 machines per day. The company makes a profit of $400 on each A machine and $600 on each B machine. How many A and B machines should the company manufacture every day to maximize its profit?

This project has two design variables:

- x_1 = number of A machines manufactured each day

- x_2 = number of B machines manufactured each day

The objective is to maximize the daily profit that can be expressed using the design variables as:

$$P = 400 * x_1 + 600 * x_2$$

The process has several design constraints:

- Shipping and handling constraint $x_1 + x_2 \leq 16$

- Manufacturing constraint $x_1/28 + x_2/14 \leq 1$

- Limitation on sale department $x_1/14 + x_2/24 \leq 1$

- Non-negative design variables $x_1 > 0, x_2 > 0$.

Using the first constraint, we can create the plain that limits the region of feasibility (Figure 8.5a).

Using the same idea and all of the constraints, a region of feasibility can be established (Figure 8.5b). Furthermore, using the region of feasibility, several contour lines can be drawn that will have the same profit, as shown in Figure 8.5c.

Finally, as the contour lines go up, they have the largest profit and the highest feasible profit line is the one that intersects with point $D(4, 12)$ with a maximum profit value of $P = 8800$.

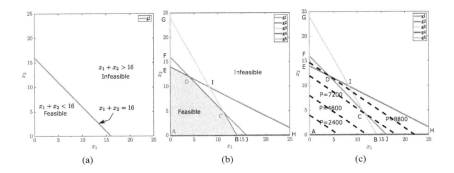

Figure 8.5 (a–c) Project 05 constraint boundary and contour lines in the profit maximization problem.

8.2 PROJECTS

8.2.1 Project 01: Linear Square Regression

Project/problem description. Using the points in Table 8.1, a regression line (calculating a_0 and a_1) that has the minimum average distance to all the points was obtained.

The left side of Figure 8.6 shows the points spread within the intervals $[0, 10]$ and $[0, 13]$. However, the right side of Figure 8.6 shows the same points and includes a regression line $y = -0.2663 + 1.3164x$, where the values of a_0 and a_1 were obtained using the least mean square regression.

TABLE 8.1 Regression Points

n	x_1	x_1
1	2.0774	3.3123
2	2.3049	3.8982
3	3.0125	4.6500
4	4.7092	6.5576
5	5.5016	7.5173
6	5.8704	7.0415
7	6.2248	7.7497
8	8.4431	11.0451
9	8.7594	9.8179
10	9.3900	12.2477

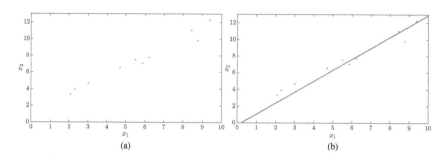

(a)   (b)

Figure 8.6 (a and b) Project 06 linear regression.

8.2.2 Project 02: Welded Cantilever Minimization

Project/problem description. A welded cantilever beam is subjected to a tip force (see Figure 8.7). The force F is applied to the cantilever beam of length L, and this force produces shear stress in the welds τ_t and τ_y. T corresponds to the torque effect of F.

The volume of the weld holding the cantilever should be as small as possible while maintaining the applied tip force. As shown in Figure 8.7, the weld has two segments of length l and height b. What is the minimum volume of weld subject to a maximum shear stress limit in the weld? The parameter values are as follows:

- Upper limit for shear stress $\tau^u = 30,000\,psi$

- Force $F = 6,000\,lb$

- Height $h = 4\,in$

- Length $L = 14\,in$

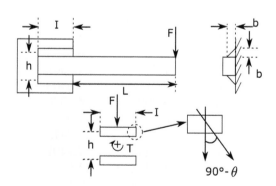

Figure 8.7 Project 08.

- Limits b: $0.125\,in \le b \le 0.75\,in$

- Limits l: $1\,in \le l \le 6\,in$

The stress shear is calculated as:

$$\tau_y = F/(bl\sqrt{2})$$
$$\tau_t = \left[6F(L + 0.5l)\sqrt{h^2 + l^2}\right] / \left[\sqrt{2}bl(l^2 + 3h^2)\right]$$
$$\tau = \sqrt{\tau_y^2 + 2\tau_y\tau_t\cos\theta + \tau_t^2}$$
$$\cos\theta = l/\sqrt{h^2 + l^2}$$

8.2.3   Project 03: Traveling Salesman Problem

Project/problem description. An example of the traveling salesman problem (TSP) with 18 cities (see appendix A.1) is given in Figure 8.8.

8.2.4   Project 04: 3D Traveling Salesman Problem

A 3D TSP application was created in LabVIEW® (see Figure 8.9). This app has the ant colony optimization (ACO) algorithm adapted to minimize the length of the tour through the cities with coordinates in 3D (x, y, z).

Figure 8.10a and b show how the user can input the coordinates with the values in X, Y, and Z axes. First, it is necessary to introduce the

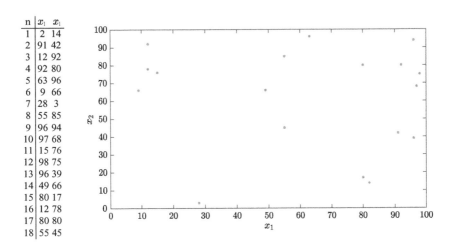

n	x_1	x_1
1	2	14
2	91	42
3	12	92
4	92	80
5	63	96
6	9	66
7	28	3
8	55	85
9	96	94
10	97	68
11	15	76
12	98	75
13	96	39
14	49	66
15	80	17
16	12	78
17	80	80
18	55	45

Figure 8.8   TSP with 18 Points

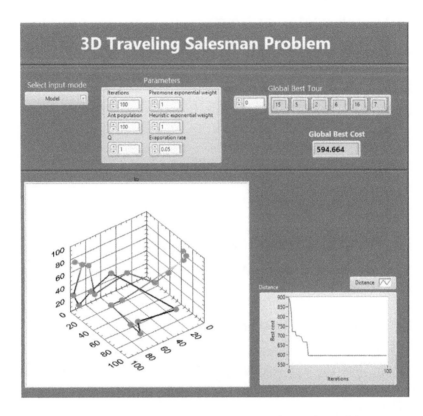

Figure 8.9   3D TSP application using a pre-configured model.

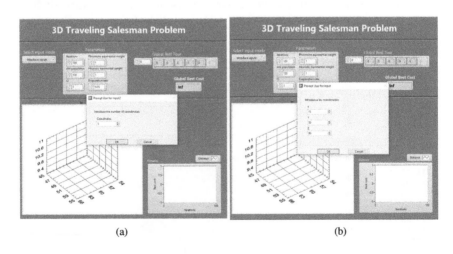

Figure 8.10   (a and b) 3D TSP application using configured coordinates by the user.

number of coordinates. Then, the user will enter the coordinates to start the optimization process.

Finally, the app shows in the graph the optimum route found by the ACO algorithm. Besides, an array shows the global best tour in which the index of the coordinate is saved in the order that the salesman must follow in order to minimize the length of the tour.

8.3 MPPT CASE STUDY

In power electronics systems, energy generation systems through the harvesting from renewable sources have gained greater relevance in an endless number of applications; among them, the photovoltaic (PV) systems have gained in popularity for plenty of solutions.

The basic scheme of a PV generator system is shown in Figure 8.11, where the PV arrays are made up of individual PV cells. The energy harvesting from the PV cells relies on the input conditions; meanwhile, the supplied energy depends on the features of the each cell and also on the configuration of the array (the number of cells connected in series and/or in parallel). As shown in Figure 8.11, i_{PV} and V_{PV} are the current and voltage supplied by the PV array, respectively, and V_o represents the voltage supplied by the power converter to a load.

Nevertheless, the main properties related to the voltage and current behaviors of a PV generator system are strongly dependent on the uncontrollable and unpredictable temperature and sun irradiation as input variables. Hence, due to the highly variable inputs, the optimal power transferring point turns into a dynamic tracking problem where the current profile of the PV cells and/or arrays interconnected with a variable

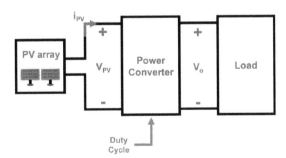

Figure 8.11 Connection scheme of a power converter linked to the PV generation system.

output impedance achieves an output voltage capable of reaching the maximum power point tracking (MPPT).

In the case of study presented in this example, Figure 8.11 shows the PV array connected to a power converter, which takes the DC energy supplied by the array and converts it into another DC output voltage, which is responsible for the output power for a certain load in the system. The power converter connected to the PV array can be a DC/DC topology where the selected structure directly depends on the supplying application.

Therefore, the scheme of Figure 8.11 can be redrawn, which is illustrated in Figure 8.12, where the power converter block is replaced by a boost DC/DC converter topology, which has the resistive load as the output of the system. Then, the relation between the changes in the duty cycle and the sum of impedances in the circuit allows us to achieve a variable resultant impedance at the output of the PV array. As shown in Figure 8.12, C_c is the coupling capacitor of the system, L the inductance, C the output capacitance, R the output load, and the MOSFET (Q) and the diode (D) are the switches that are in charge of the commutation between the states of the system, respectively.

Then, the dynamic optimization issue can be dealt with different MPPT solutions. One of the most implemented algorithms is the classic Perturb and Observe (P&O), which is widely used due its simplicity although it does not usually get the fastest response and mostly has more variations when approaching the MPP. The P&O algorithm is based on the injection of small perturbations into the system, whose effects are observed in order to direct the operating point in the direction of the MPP.

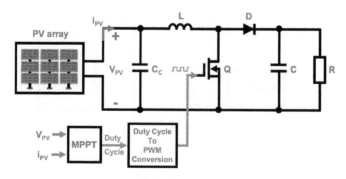

Figure 8.12 Selected topology for the case study.

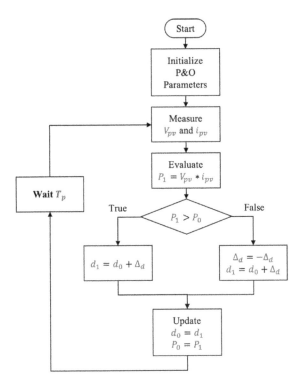

Figure 8.13   Flowchart of the P&O MPPT algorithm.

Therefore, to understand the optimizing behavior of the technique, Figure 8.13 shows the basic flowchart of the P&O MPPT algorithm, which basically starts with the system in a determined operating point, which is perturbed every T_P period of time, by changing the duty cycle d in order to apply a voltage variation to the PV source terminals. After each applied disturbance, the algorithm seeks to compare the resultant power supplied by the PV array (P_{PV}), where the change in the measured power (P_{PV_1}) compared to its previous power value (P_{PV_0}) leads to the direction of the next perturbation to be applied.

Moreover, as explained in [3], the P&O MPPT algorithm has been the seminal solution that enabled plenty derivations for MPPT applications, such as the hill climbing (HC) method and metaheuristic-based approaches. Thus, an MPPT method based on the particle swarm optimization (PSO) algorithm has been proposed by [3,19], where the online optimization is performed by perturbations applied to the system based on the classical PSO velocity and position equations already presented in this book.

In this particular case study, the positions given by x are the duty cycles to be uploaded into the system, and the velocities (v) are the variations to be uploaded into the j positions, in order to achieve the perturbative behavior. Then, Figure 8.13 shows the flowchart of the PSO-based MPPT algorithm, where G_{best} is the MPP and x_{best} the value of the duty cycle where it was achieved. Additionally, it is important to highlight that MPPT approaches based on metaheuristic methods retain most of the virtues and hence the defects of the original metaheuristic optimization algorithms.

For the implementation of the PSO to the MPPT, some considerations have to be made in order to adapt the algorithm to the dynamic search: first, the initialization of the PSO parameters, such as the dampening coefficients, the global and particular learning coefficients, the velocities (v), and positions (x) of the population of particles.

Thereafter, Figure 8.14 shows the general structure of the PSO adapted as an MPPT algorithm: After the initialization of the PSO parameters, the algorithm begins with the measurement of the voltage and the current from the PV source. Then, this process has two threads depending on whether the algorithm is loading the particles to the duty cycle or not:

- When the counter j differs from the number of particles N_P, the position of each particle is charged one by one to the duty cycle, just as in the P&O algorithm.

- On the other hand, when the counter j matches the number of particles N_P, the method takes the natural behavior of the typical PSO by obtaining and updating the MPP and the best positions.

Additionally in the case where $j == N_P$, a scanning process has to be made in order to detect irradiation changes, a task that can be achieved by comparing the previous MPP value against the new one, which determines if the system still achieves the saved maximum power transfer with the duty cycle saved as the current best position. If a change in solar irradiation is detected, then the PSO parameters must be reinitialized to ensure that the algorithm does not stagnate in a local solution.

Moreover, for this example, another metaheuristic-based MPPT is presented where the proposal includes a version of the earthquake algorithm (EA) adapted for the task. The main structure of the adaptation of the geo-inspired algorithm is shown in Figure 8.15.

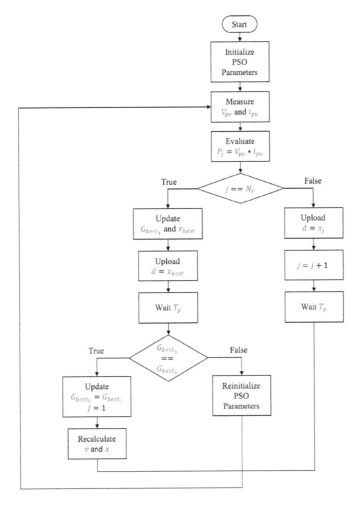

Figure 8.14 Flowchart of the PSO MPPT algorithm.

As explained in Ref. [45], the EA within its *S-range* (Sr) has a fine searching behavior, characterized for being around the global best value. Thus, for extrapolating that feature for the MPPT dynamic optimization, the solution requires contemplating that the epicenters should be "orbiting" around the MPP, in order to ensure a correct optimizing behavior.

Under those circumstances, a searching flag $(S_f lag)$ is implemented to intercalate the best duty cycle within the searching positions, achieve the expected routeways during the searching stage, and also extend the time that the system will remain in the considered MPP.

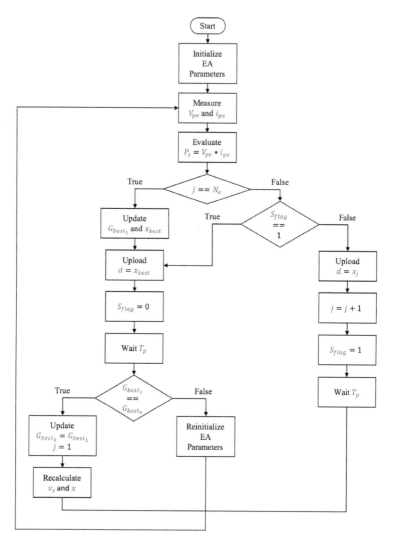

Figure 8.15 Flowchart of the EA MPPT algorithm.

Unlike the PSO approach, the original EA has two velocity equations: This is because for its first implementation, it is important to highlight that the v_s equation is taken into account.

8.3.1 Simulink Models

The Simulink® model shown in Figure 8.16 shows how the simulations were performed using the *Specialized Power Systems* components from the *Simscape* library. Thus, the boost converters are implemented, which

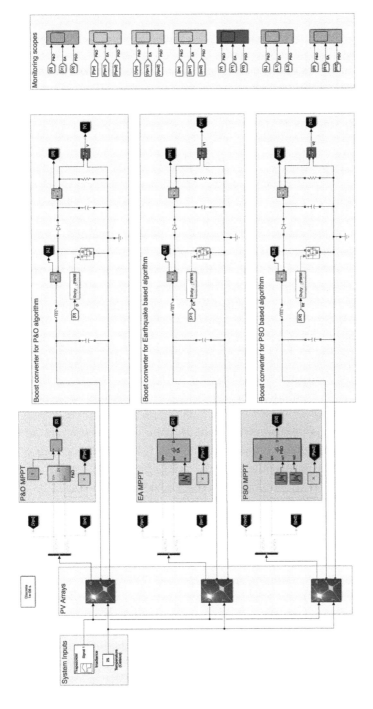

Figure 8.16 Flowchart of the EA MPPT algorithm.

are highlighted in gray color; meanwhile, the light blue groups integrate the components that are necessary for the MPPT implementations, and the orange blocks are the random variables taken by the metaheuristic algorithms.

Additionally as shown in Figure 8.16, the green blocks are the voltage sensors and the yellow ones, current sensors. The elements interconnected with magenta lines are the voltages, currents, and resultant powers measured from the PV arrays (highlighted in the light green box).

Grouped in the light red box, the monitoring scopes are implemented in order to compare the different performances, where each graph complies with the color code referring to the components of the model.

The boost converter for the simulations was designed to operate in CCM (continuous conduction mode) for a typical 12[V] input voltage conversion into 24[V] output voltage at 2[A]. A summary of the designed boost converter is shown in Table 8.2, where the nominal values of operation such as the switching frequency, input and output voltages, the output power, and current are presented.

Meanwhile, the coupling capacitor (C_c as shown in Figure 8.12) is used for the interconnection with the PV arrays; for each converter case, a 1000[μF] capacitance was selected. On the other hand, the parameters of the PV array configuration selected for the simulations are summarized in Table 8.3.

Finally, the simulations were made using the parameters as shown in Table 8.4, where the simulation mode is taken as discrete in order to achieve a correct emulation with the sampling time.

In the case of dynamic system simulations, equations that involve one or more derivatives of a dependent variable with respect to a

TABLE 8.2 Converter Parameters

Input voltage	12 [V]
Switching frequency	120 [kHz]
Duty cycle	50 [%]
Inductor	47 [mH]
Capacitor	220 [μF]
Load resistance	12 [Ω]
Output power	48 [W]
Output voltage	24 [V]
Output current	2 [A]

TABLE 8.3 PV Array for Low-Power Simulation Parameters

Module	CRM60S125S
Maximum power	57.96 [W]
Cells per module	24
Open-circuit voltage (V_{oc})	14.5 [V]
Short-circuit current (I_{sc})	5.51 [A]
Voltage at MPP (V_{mp}) (V)	11.5 [V]
Current at MPP (I_{mp})	5.04 [A]
Temperature coefficient of V_{oc}	−0.322 [%/°C]
Temperature coefficient of I_{sc}	0.071996 [%/°C]
Light-generated current (I_L)	5.5363 [A]
Diode saturation current (I_O)	4.6762e-11 [A]
Diode ideality factor	0.92399
Shunt resistance (R_{sh})	56.6705 [Ω]
Series resistance (R_s)	0.25984 [Ω]
Parallel strings	1
Series-connected strings	1

TABLE 8.4 Simulations Parameters Used in MATLAB® Simulink®

Simulation Mode	Discrete
Stop time	10 [s]
Sampling time	1 [μs]
Time to start saving data	0 [s]
Solver	ode4 (Runge–Kutta)

single independent variable, are known as ordinary differential equation (ODE). Consequently, MATLAB®'s Simulink uses different kind of solvers in order to iteratively find solutions to the ODE, in different time lapses. Therefore, for this case study, *ode4 (Runge–Kutta)* is selected since it is a fixed-step solver that allows us to make a numerical method approximation [47].

8.3.2 Results

The implementation of the algorithms is shown in Figures 8.13–8.15, whereas the analysis of simulated behavior of the three algorithms is shown in Figure 8.17.

The plot in Figure 8.17 shows the dynamic optimization profiles obtained from the PV power P_{pv} (highlighted in Fig 8.16), where the P&O

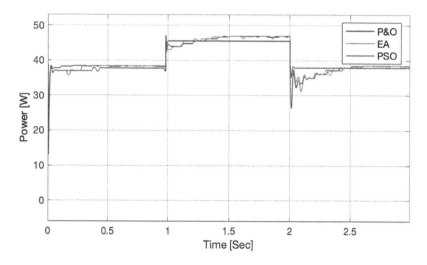

Figure 8.17 MPPT through time.

implementation is shown in green, and the EA and PSO approaches in red and blue, respectively.

Therefore, the P&O algorithm was configured using a 0.01 step size every 0.1 seconds; meanwhile, the EA was calibrated using 2 epicenters, and the PSO tuned using an inertia weight $w = 0.85$, 1.25 for the personal and global learning coefficients and 4 particles for the optimization process.

On the one hand, the PSO approach was highly dependent on the correct calibration of its parameters, with a great performance in terms of steady-state oscillations. Additionally, it showed a fast response against irradiation changes, with a drawback in the accuracy to find with precision the MPP.

On the other hand, the EA-based MPPT algorithm showed a better dynamic tracking behavior; nevertheless, the main drawback of the algorithm against the PSO is the steady-state oscillation. However, the consistency of the algorithm against different performed tests showed a low dependence on the calibration of the algorithm achieving greater reliability.

Both metaheuristic-based algorithms were compared with the traditional P&O algorithm, in order to verify their behavior and the MPPT against different irradiances, validating that in both cases, the dynamic optimization process can be done through the adaptation of metaheuristic algorithms.

8.4   INDUSTRY 4.0 CASE STUDY: THREE-PHASE INVERTER

Today, the requirement of developing new products is extremely high since the end users demand those. Moreover, every year-end consumers are waiting for new products. If a company is not developing new product, it cannot be a technological reference company.

New advances in technology have made a significant growth in the industrial productivity. For example, factories were powered by steam engine in the 19th century, electricity allowed mass production in the 20th century, and automation helped industries in the 1970s. The new form of advancement is Industry 4.0, which is managed by nine pillars [69]: autonomous robots, simulation, horizontal and vertical system integration, the industrial internet of things, cybersecurity, the cloud, additive manufacturing, augmented reality, and big data and analytics.

Oriented toward smart manufacturing implementations, a three-phase inverter case study is presented. Inverters in power electronics have been widely used for industrial power applications and home appliances, such as constant voltage and frequency power supplies, PV generation system connection to the grid, and control of AC and brushless direct current (BLDC) motors, among other applications (Figure 8.18).

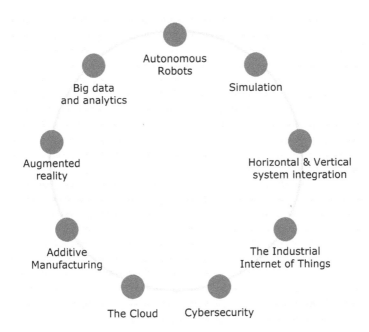

Figure 8.18   Industry 4.0.

TABLE 8.5   Design Parameters of the Inverter

Maximum power	1600 [W]
DC bus voltage	12 - 100 [V]
Logic voltage	1.2 - 6 [V]
Switching frequency	120 [kHz]
Turn-on propagation delay	125 [ns]
Turn-off propagation delay	105 [ns]

The designed inverter in this case study has its main objective, that is, to serve as a driver that allows us to control BLDC motors. The most relevant parameters of the implemented circuit are summarized in Table 8.5.

Since the design was implemented as an early testbed iteration of a surface mounting circuit, the selected components are dual in-line packages (DIP), which for instance can be through-hole-mounted or inserted in a socket in a printed circuit board (PCB) (as in this case study).

Then, PCBs manufactured with DIP components often require different diameters of perforations, due to the variety of diameters on the pins of the different components. Hence, computer numerical control (CNC) machines intended for PCB manufacturing require tools of a wide variety of diameters, which can be changed depending on the required drills in the designs.

Nevertheless, trajectory efficiency in conjunction with the time lost in tool changes often compromises the manufacturing time and in some cases the structural integrity of the tools, because the optimization of the time that the tools are used, if this time is decreased, the lifetime of the machine could be increased and the process time could be reduced.

Figure 8.19a and 8.19b show the simulated diameters and trajectories of the design without the optimization, respectively, where the copper plate used for the manufacturing process is 20[mm] wide by 20[mm] long. As shown in Figure 8.19a, the design requires drills with six different tool diameters.

The 0.8128[mm] drill holes are required by the sockets for the components with DIP packages, such as the microcontroller ATmega 16, the MOSFET gate drivers IR2112, and the serial interface for the microcontroller MAX232, among others. Additionally, some of the through-hole mounting components such as resistances, capacitors, and the crystal oscillator have 3.302[mm] drilling diameter. The most commonly required drilling radius in the design is shown in Figure 8.19a.

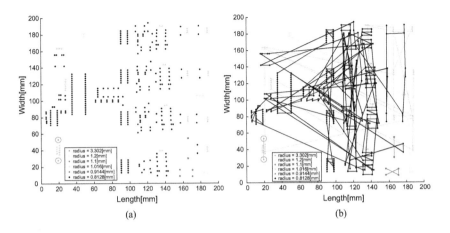

Figure 8.19 (a and b) Simulated PCB.

On the other hand, the 1.2[mm] drill holes are required by the selected screw terminals, which are used for the digital power connections, the motor connections, and the voltage of the power electronic stage.

The 1.1[mm] drilling radius is only required by the 5[W]–10[Ω] resistance selected for the dynamic brake stage. Meanwhile, the 1.016[mm] drills are required for the terminals of the DV-9 female terminal (for USB-Serial communication), the headers, the connectors, the linear voltage regulator 7805, and the IRF3710 MOSFETs. Nevertheless, the 3.302[mm] drilling holes are required by the DV-9 female terminal, which is used to ensure a good connection with the USB-Serial cable.

Finally, the 0.9144[mm] drill holes are necessary for the 104 K polyester capacitors, connected in parallel to the power-stage supply. Table 8.6 summarizes all the components needed for the design, in addition to the drilling holes required for their mounting, the number of mounting holes required per package, and the quantity of packages used for each component.

Meanwhile, Figure 8.19b shows the trajectories of every tool, according to the coordinate sequence, sent by the PCB software used for the designing task (EAGLE). Table 8.7 shows the summary of the drills by each tool radius required in the design.

Then, as it is shown in Figure 8.19b and later discussed in Table 8.7, the trajectory optimization issue to be solved in this case study is related to the 361 drilling points, considering the required six different tool radii in the design.

TABLE 8.6   Selected Components for the Design

Qty.	Component	Drill Radius [mm]	Required Drills
1	MAX232	0.8128	16
1	ATmega16	0.8128	40
4	IR2112 gate drivers	0.8128	14
7	IRF3710 MOSFETs	1.016	3
1	Crystal oscillator	0.8128	14
9	Electrolytic capacitors	0.8128	2
23	Ceramic capacitors	0.8128	2
2	Polyester capacitors	0.9144	2
18	1N4148 Diodes	0.8128	2
1	7805	1.016	3
1	5[W]-10[Ω] resistor	1.1	2
26	Resistors	0.8128	2
1	DV-9 female connector	1.016	9
		3.302	2
1	3 pin screw terminal	1.2	3
2	2 pin screw terminal	1.2	2
19	Male headers	1.2	1

TABLE 8.7   Required Drills

Drill Radius [mm]	Required Drills
3.302	2
1.2	7
1.1	2
1.016	47
0.9144	4
0.8128	299
Total	**361**

8.4.1   MATLAB Optimization Solution

In this problem, it is required not only to consider the distances between each perforation, but also to consider that each time there is tool change and some extra time is spent. To solve this TSP problem, the distance for each perforation was included and adapted so that it includes a penalization for tool changing. Including the penalization adapts the problem can take into consideration if a tool moves an additional distance the process time is increased at each end-point of the process.

To optimize the perforation distance and the tool change, the ACO algorithm was chosen for each drilling point. The algorithm was run over

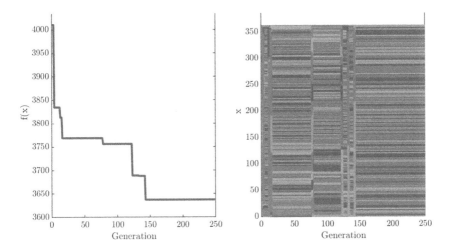

Figure 8.20 Evolution soldering. This image shows the evolution of how the distance was reduced for each iteration using ACO to reduce the drilling distances and tool changes.

250 iterations with a population of 30 ants and learning parameters such as rate of evaporation $\rho = 0.5$, control of influence $\alpha = 2$, control parameter $\beta = 6$, and pheromone deposited constant $Q = 1$. The evolution of the training is shown in Figure 8.20, where the image on the left represents the distance evolution, which was reduced to 3640 mm (including the penalization) and the image on the right represents how each point changes the position in the algorithm.

Additionally, Figure 8.21 shows the final path for the drilling process. Each color represents the final drilling process with each of the tools. The red line represents a moment when it is required to change tools between drilling. It is worth noticing that using this penalization, it keeps moving from hole to hole with the same diameter for almost all the drilling process, and that only in a rare occasion, it changes tools before continuing with drilling the same diameter holes (the yellow line that splits drilling process into two). The final total number of times it changes tools was eight.

8.4.2 LabVIEW Optimization Solution

For implementation, an application in LabVIEW was developed. This app has the ACO algorithm embedded to minimize the final path of the drilling process for the inverter PCB. First, a file (.txt) is read with the

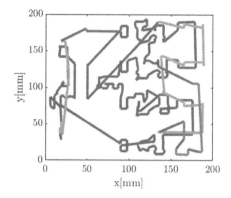

Figure 8.21 Final path for the drilling process. Each line represents the path taken for each of the different hole diameters, and the dark gray line represents a change of tool between the paths. It can be seen that for the most of the drilling processes, it finishes to drill one diameter and moves to the next diameter to find the shortest distance. This changes for the light gray where it separates the drilling into two sections and changes tool before continuing with that diameter.

coordinates (in X and Y) and the tool is necessary for each drilling point. After that, a model is created in which the distance between coordinates is calculated. Then, the ACO algorithm starts to reduce the drilling distances. Figure 8.22 shows the final result of the optimization process. The final path and the tool changes are visualized in the graph. The total length is reduced to 2874 mm, and the tool changes are 10 times.

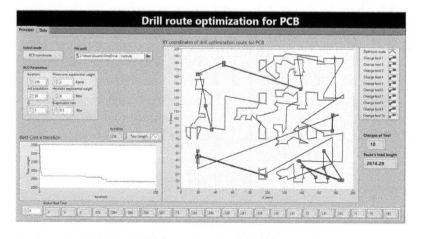

Figure 8.22 Drill route optimization application.

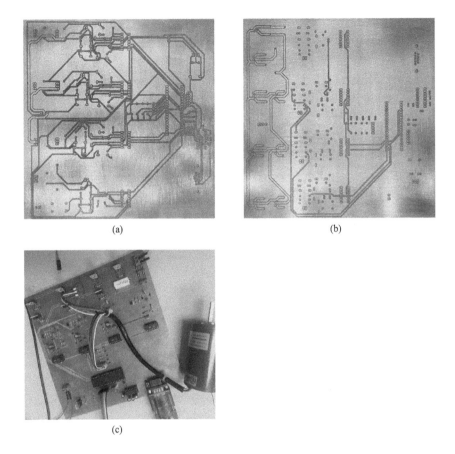

(a)

(b)

(c)

Figure 8.23 Final PCB views: a) top and b) bottom view without components; c) PCB with motor and cable connections.

8.4.3 Final PCB

The final PCB is shown in Figure 8.23, where the first row shows a) the top and b) bottom view without components, and the bottom row shows c) the PCB with the motor and cable connections.

8.5 DC MOTOR SPEED CONTROLLER WITH PID TUNING OPTIMIZATION ALGORITHM

This section presents the application: DC motor speed controller with proportional-integral-derivative (PID) tuning optimization algorithms. This application was implemented in a FPGA cRIO-9030 of National Instruments[TM] using LabVIEW as a visual programming language and

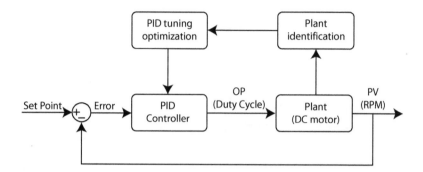

Figure 8.24 Block diagram of control system with PID tuning optimization.

interface. The controller used for this case study was an embedded PID. In addition, a PID tuning optimization system was set up in order to find the optimum PID gains to improve the transient response of the system. Figure 8.24 shows the block diagram of the control system, in which the configured optimization algorithms used were as follows: (1) PSO, (2) BA (bat algorithm), (3) EA, (4) GWO (gray wolf optimization), and (5) ACO.

To evaluate and compare the improvements of the optimization algorithm's implementation, performance indices were selected to measure objective functions of the dynamic and stationary errors [29]. The improvement of the implementation was analyzed using four performance indices, which are as follows:

1. The integral of the square of the error, *ISE*, can be written as

$$ISE = \int_0^T e^2(t)dt \tag{8.1}$$

2. The integral of the absolute magnitude of the error, *IAE*, is defined as

$$IAE = \int_0^T |e(t)|dt \tag{8.2}$$

3. The integral of time multiplied by the absolute error, *ITAE*, is defined as

$$ITAE = \int_0^T t \cdot |e(t)|dt \tag{8.3}$$

4. The integral of time multiplied by the squared error, *ITSE*, can be written as

$$ITSE = \int_0^T t \cdot e^2(t)dt. \tag{8.4}$$

Figure 8.25 presents the general topology of the system, which is composed of three principal elements: (1) plant (DC motor), (2) embedded controller (cRIO-9030 FPGA), and (3) optimization system (PC host).

The plant consists of a DC motor with a 5V of nominal voltage controlled by an H-bridge, which is manipulated by a pulse width modulation (PWM) signal. Additionally, a feedback system made by a disk with a notch with an optical sensor is created in order to measure the speed of the motor (RPM) (see Appendix A.2 for detailed information).

The PID controller was embedded into the cRIO FPGA, and a module NI 9401 was used as a digital input/output (I/O) for data acquisition card. In addition, two channels of first in first out (FIFO) were created to

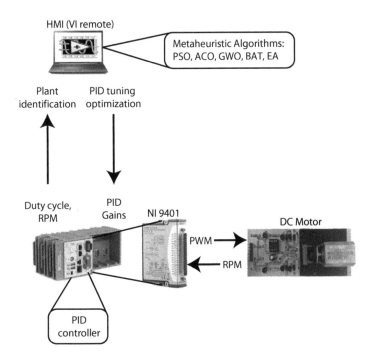

Figure 8.25 General structure of control system with PID tuning optimization.

send data of manipulation (PWM) and feedback (RPM) of the plant to the PC host where the optimization algorithms were implemented. The PC host receives the process variable (RPM) and manipulation (PWM) signals from the FPGA (field-programmable gate array) to identify the parameters of the plant. The sub-VI used to identify the parameters of an autoregressive-moving-average (ARX) model was the *SI Recursively ARX model* (*SISO Waveform*). The *normalized least mean squares* (*NLMS*) was the recursive estimation method used to obtain the plant model in second order, as follows:

$$G(z) = \frac{Y(z)}{U(z)} = \frac{5.834 + 5.865z^{-1}}{1 - 0.564z^{-1} - 0.325z^{-2}}. \qquad (8.5)$$

The PC host consists of a LabVIEW App which is communicating with the cRIO FPGA where the control system is embedded. Figure 8.26 shows the front panel of the application named "DC motor speed controller with PID tuning optimization algorithms".

Before starting optimization process, it is necessary to identify the plant through the manual mode. Figure 8.27 presents the front panel of the app in manual mode. A step in manipulation (0%–100% in duty cycle) is introduced to generate a response of the plant in order to find the parameters of the plant model (Equation 8.5).

Once the identification is performed, the PID tuning optimization process can be realized, and the user has to select any of the aforementioned optimization algorithms. It is required to introduce the parameters based on each algorithm. The objective function is estimated from

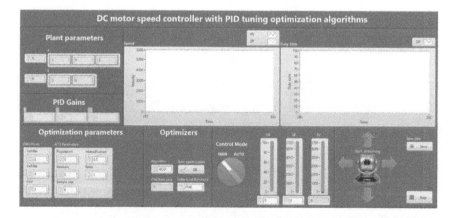

Figure 8.26 DC motor speed controller (front panel).

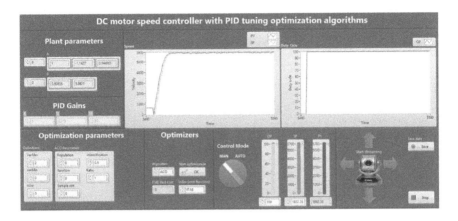

Figure 8.27 DC motor speed controller (manual mode for plant identification).

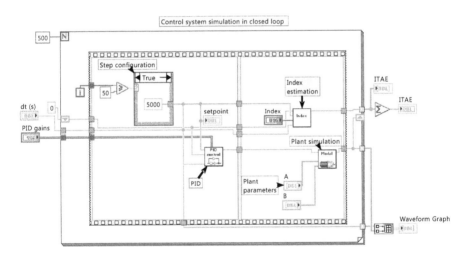

Figure 8.28 Simulation of control system in closed loop.

the transient response in simulation of the control system in closed loop. Figure 8.28 shows the LabVIEW implementation for objective function estimation; this sub-VI simulates the control system with the plant parameters found in the identification process.

Figure 8.29 shows the app in automatic mode with the PID gains obtained from the optimization process.

In addition, a video streaming was implemented to get remote control (see Figure 8.30).

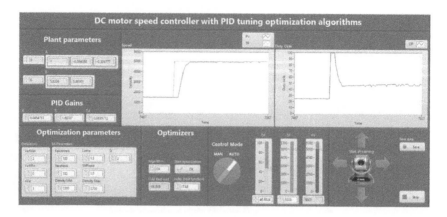

Figure 8.29 DC motor speed controller (automatic mode with PID gains obtained by metaheuristic algorithms).

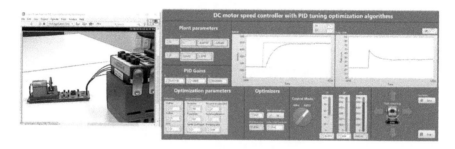

Figure 8.30 DC motor speed controller (remote application and monitoring by video streaming).

8.6 OPTIMIZATION ALGORITHMS EMBEDDED IN LABVIEW FPGA

Field-programmable gate arrays (*FPGAs*) have become a multi-market solution. The scalability, energy efficiency, and the capacity of handling great volumes of streaming data are properties that make FPGA more potent compared with traditional microprocessors due to its parallel processing that increases its performance per watt [31]. Hence, this has made it possible to develop new solutions for engineering applications. For example, a customized FPGA called Catapult has been designed by Microsoft™ for its data center, which improved the ranking throughput of the Bing search engine by 2x [9].

The aforementioned properties introduce an opportunity to implement optimization algorithms in order to optimize processes in the same

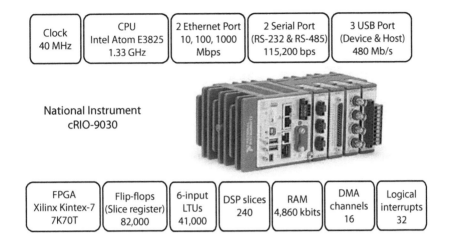

Figure 8.31 National Instruments cRIO-9030 FPGA specifications.

way as online or parallel in real time. Based on these properties, five metaheuristic algorithms were implemented in a FPGA *cRIO*-9030 of National Instruments^{TM} (*NI*) in order to analyze their performance and computational cost. A set of standard benchmark functions is selected to make a comparison between the algorithms. The metaheuristic algorithms implemented in a FPGA were PSO, BA, GWO, EA, and NM (Nelder–Mead).

The principal specifications of the FPGA, where the optimization algorithms were implemented, are mentioned in Figure 8.31. The NI cRIO-9030 is an embedded controller with real-time processor and reconfigurable FPGA, ideal for monitoring and controlling applications in real time. The main characteristics of the FPGA are as follows:

- FPGA type: Xilinx Kintex-7 7K70T

- Clock: 40 MHz

- Flip-flops (slice registers): 82,000

- 6-input Look-up tables (*LUTs*): 41,000

- Digital signal processing (*DSP*) slices: 240

- Random access memory (*RAM*): 4,860 kbits

- Direct memory access (*DMA*) channels: 16

- Logical interrupts: 32.

8.6.1 Benchmark Functions

To evaluate the proposed optimization methods, the performance of the algorithms is tested in this section using ten benchmark functions.

Standard test functions are used to evaluate the optimization algorithms; the benchmark functions selected were as diverse as possible to analyze how the algorithms perform between different conditions, for example, in a convergence problem like the convex *spheric function*, against unimodal and non-convex functions like the *Ackley function*, or even under multimodal non-convex functions with several local or global solutions, like the *Hölder table function*.

The surfaces of the benchmark functions used are shown in Figure 8.32. Additionally, the mathematical representation for each function is given in Table 8.8, where the global solutions and the search domains for each function are also shown. Those parameters were used as the benchmark for the optimization methods.

To compare the optimization methods, every algorithm was implemented with a population of 15 search agents. Also, it is important to mention that the results of the algorithms were evaluated after 100 iterations, and each was repeated 500 times.

8.6.1.1 *Implementation into FPGA*

To implement some benchmark functions into the FPGA, it was necessary to develop a sub-VIs with trigonometric functions (*cos* & *sin*) in order to reduce the operation cost through the use of a memory as a 2D-LUT. Figure 8.33 shows how a $\pi/2$ cycle of *sin* function was implemented into a memory using 1024 slots.

Figures 8.34 and 8.35 depict the programmed sub-VIs to calculate the *sin* and *cos* functions, respectively. First, Equation 8.6 was implemented in both sub-VIs to recalculate the x value into the 2π cycle. Then, a classifier is used to identify in which quarter of cycle between $0 - 2\pi$, the x is located.

$$f(x) = |x| - floor\left(\frac{|x|}{2\pi}\right) \tag{8.6}$$

The equations in 8.7 were implemented in the case structure as shown in Figure 8.34 to find the address that represents $f(x) = sin(x)$. Next, this value of $sin(x)$ was negated depending on the sign of x or the negative parts of function's cycle.

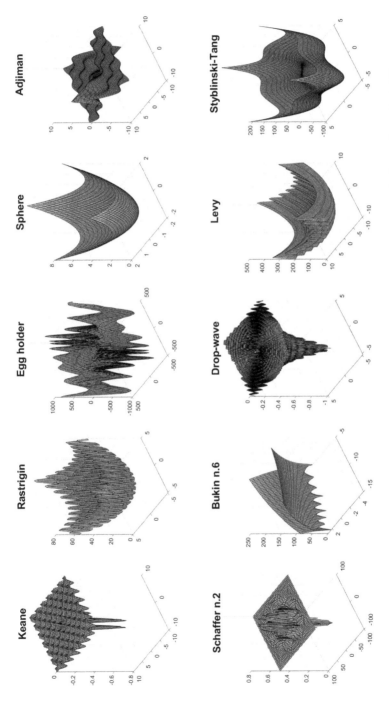

Figure 8.32 Surfaces of the benchmark functions used to compare the performance of algorithms [34;48].

TABLE 8.8 Benchmark Functions Implemented in the FPGA Used to Evaluate Each Optimization Algorithm

Function	Formula	Global Minimum	Search Domain
Keane	$f(x,y) = -\dfrac{sin^2(x-y)\,sin^2(x+y)}{\sqrt{x^2+y^2}}$	$\text{Min} = \begin{cases} f(1.393, 0) & = 0.674 \\ f(0, 1.393) & = 0.674 \end{cases}$	$-10 \leq x, y \leq 10$
Rastrigin	$f(x) = 10d + \displaystyle\sum_{i=1}^{d}[x_i^2 - 10cos(2\pi x_i)]$ where $d = 2$	$f(0,0) = 0$	$-5.12 \leq x_i \leq 5.12$
Egg Holder	$f(x,y) = -(y+47)sin\left(\sqrt{\left\|y+\dfrac{x}{2}+47\right\|}\right)$ $\quad - x sin\left(\sqrt{\|x-(y+47)\|}\right)$	$f(512, 404.232) = -959.6407$	$-512 \leq x, y \leq 512$
Sphere	$f(x) = \displaystyle\sum_{i=1}^{d} x_i^2$	$f(0,0) = 0$	$-\infty \leq x_i \leq \infty$
Adjiman	$f(x,y) = cos(x)sin(x) - \dfrac{x}{y^2+1}$	$f(0,0) = -2.02181$	$-10 \leq x, y \leq 10$
Schaffer N. 2	$f(x,y) = 0.5 + \dfrac{sin^2(x^2 - y^2) - 0.5}{[1 + 0.001(x^2 + y^2)]^2}$	$f(0,0) = 0$	$-100 \leq x, y \leq 100$

(Continued)

TABLE 8.8 (*Continued*) Benchmark Functions Implemented in the FPGA Used to Evaluate Each Optimization Algorithm

Function	Formula	Global Minimum	Search Domain
Bukin n.6	$f(x,y) = 100\sqrt{(\|y - 0.01x^2\|)} + 0.01\|x + 10\|$	$f(-10, 1) = 0$	$-15 \leq x \leq -5,$ $-3 \leq y \leq 3$
Drop-wave	$f(x,y) = -\dfrac{1 + cos\left(12\sqrt{x^2 + y^2}\right)}{(0.5(x^2 + y^2) + 2)}$	$f(0, 0) = -1$	$-5.2 \leq x, y \leq 5.2$
Levy N. 13	$f(x,y) = sin^2 3\pi x + (x - 1)^2 (1 + sin^2 3\pi y)$ $+ (y - 1)^2 (1 + sin^2 2\pi y)$	$f(1, 1) = 0$	$-10 \leq x, y \leq 10$
Styblinski-Tang	$f(x) = \dfrac{1}{2}\sum_{i=1}^{d}(x_i^4 - 16x_i^2 + 5x_i)$ where $d = 2$	$f(-2.9, -2.9) = -39.166d$	$-5 \leq x_i \leq 5$

Address	Value
0	0
1	0.00153398
2	0.00306796
⋮	⋮
1021	0.999989
1022	0.999995
1023	0.999999

Figure 8.33 Memory used as a 2D-LUT.

$$Address = \begin{cases} \text{if } 0 \le x \le \pi/2 & \text{then } f(x) = \dfrac{2048x}{\pi} \\[2mm] \text{if } \pi/2 > x \le \pi & \text{then } f(x) = 2048 - \dfrac{2048x}{\pi} \\[2mm] \text{if } \pi > x \le 3\pi/2 & \text{then } f(x) = \dfrac{2048x}{\pi} - 2048 \\[2mm] \text{if } 3\pi/2 > x \le 2\pi & \text{then } f(x) = 4096 - \dfrac{2048x}{\pi} \end{cases} \quad (8.7)$$

As similar way, equations in 8.8 were implemented in the case structure as shown in Figure 8.35 to find the address that represents $f(x) = cos(x)$. After that, the value of $cos(x)$ was negated depending on the sign of x or the negative parts of function's cycle.

$$Address = \begin{cases} \text{if } 0 \le x \le \pi/2 & \text{then } f(x) = 2048 - \dfrac{2048x}{\pi} \\[2mm] \text{if } \pi/2 > x \le \pi & \text{then } f(x) = \dfrac{2048x}{\pi} - 2048 \\[2mm] \text{if } \pi > x \le 3\pi/2 & \text{then } f(x) = 4096 - \dfrac{2048x}{\pi} \\[2mm] \text{if } 3\pi/2 > x \le 2\pi & \text{then } f(x) = \dfrac{2048x}{\pi} \end{cases} \quad (8.8)$$

Once implemented the sub-VIs of trigonometric functions (cos & sin), the sub-VIs of benchmark functions were programmed and embedded. Figure 8.36 shows how the Keane function was programmed.

8.6.1.2 Benchmark Functions Utilization Summary

Table 8.9 shows the summary of the utilization by each benchmark function embedded into the FPGA.

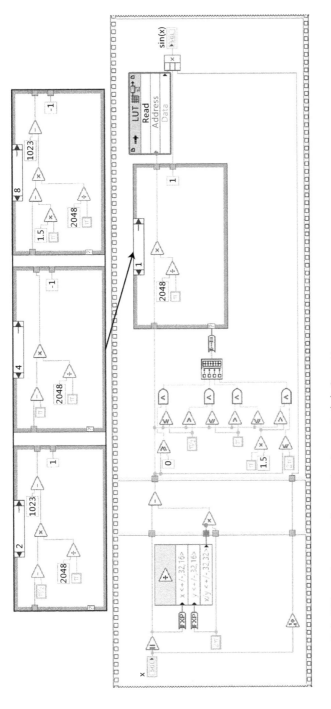

Figure 8.34 Sub-VI implemented to calculate $sin(x)$ function.

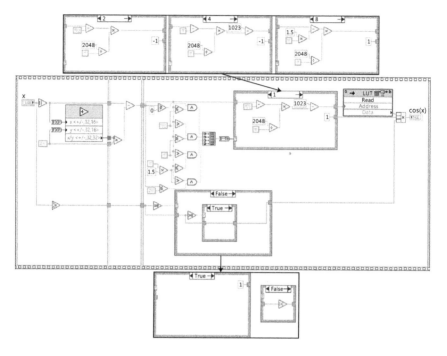

Figure 8.35 Sub-VI implemented to calculate $cos(x)$ function.

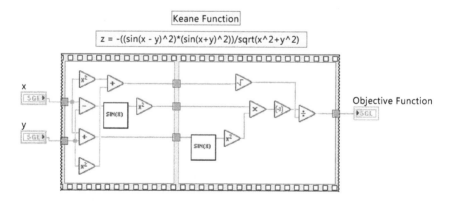

Figure 8.36 VI of Keane function.

8.6.2 Optimization Algorithms Implementation

The optimization algorithms implementation was composed of three principal elements (VIs): (1) objective function (benchmark function), (2) pseudorandom number generation, and (3) optimization algorithm.

TABLE 8.9 FPGA Utilization Summary with Benchmark Functions

Function	Slice Registers	Slice LUTs	Block RAMs	DSP48s
Keane	13,826	18,480	4	25
Rastrigin	14,690	20,563	4	25
Egg holder	8,912	11,411	4	12
Sphere	2,835	2,676	3	2
Adjiman	13,527	18,328	4	22
Schaffer n.2	9,034	12,073	4	15
Bukin n.6	3,749	3,849	3	4
Drop-wave	9,297	12,232	4	14
Levy	19,740	27,372	4	42
Styblinski–Tang	5,355	6,940	3	12

In order to generate random numbers, an algorithm described as follows was used.

8.6.2.1 Pseudo Random Number Generation

To implement the optimization algorithms, a sub-VI was compiled to generate pseudorandom numbers using a linear feedback shift register algorithm [26]. This sub-VI was built as an IP core LV FPGA, and it was found in the National Instruments[TM] community support [4]. Figure 8.37 shows the block diagram of the VI.

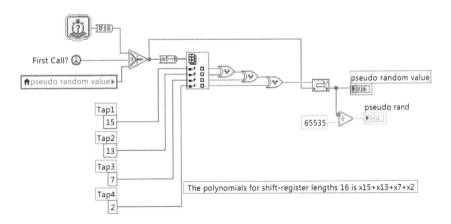

Figure 8.37 Sub-VI of pseudorandom number generation.

8.6.3 Optimization Algorithms Utilization Summary

Table 8.10 and Figure 8.38 show the device utilization summary from each optimization algorithm embedded into the FPGA. Besides, Table 8.9 indicates the ticks used per algorithm in process optimization (1 tick = 40 MHz). For example, in PSO algorithm, the optimization process cost was 148, 621 ticks = 3.715 ms.

TABLE 8.10   FPGA Utilization Summary with Optimization Algorithms (Sphere as Objective Function)

Algorithm	Slice Registers	Slice LUTs	Block RAMs	DSP48s	Ticks
PSO	40,504	31,827	3	17	148,621
BA	38,993	29,967	3	10	231,953
GWO	27,568	28,369	3	26	797,735
EA	42,489	33,923	3	21	712,631
NM	27,204	33,164	3	34	27,313

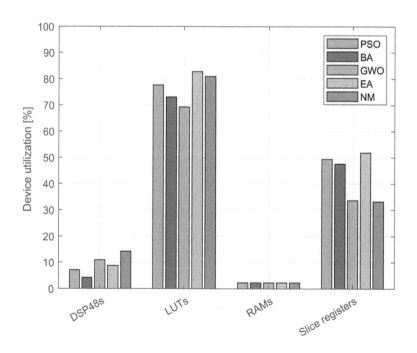

Figure 8.38   Device utilization summary (%).

Appendix

A.1 TRAVELING SALESMAN PERSON

A common optimization problem is the traveling salesman problem. This problem estimates that given a group of cities with a certain cost to travel through them or distance, what is the cheapest way or the shortest route to visit all of the cities. One simple cost function for the problem is just the distance traveled by the salesperson for the given (x_n, y_n), for cities $n = 1, ..., N$ given by

$$c = \sum_{n=0}^{N} \sqrt{(x_n - x_{n+1})^2 + (y_n - y_{n+1})^2} \qquad (A.1)$$

A simple example is a problem with N = 13 cities given the start and end points that the obvious best configuration is where all the cities lie in a rectangle with a minimum distance of 14 (Figure A.1).

A.2 DC MOTOR IMPLEMENTATION

The experimental system designed for the tests consisted of a DC motor with a 5 V of nominal voltage controlled by an H-bridge, like the one shown in Figure A.2, which also shows the resultant circuit after activating the transistor Q_5 where the red segmented line represents the activation current flow and the green line the resultant energy leading to the motor movement.

Additionally, in order to ensure the quality of the DC motor's measured *rotations per minute* (RPM), a test environment was designed where the motor could be fixed to a base with a disk having a notch

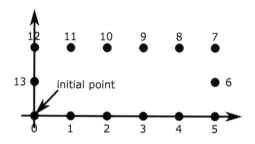

Figure A.1 Graph of 13 cities arranged in a rectangle. The salesperson starts at the origin and visits all 13 cities once and returns to the starting point. The obvious solution is to trace the rectangle, which has a distance of 14.

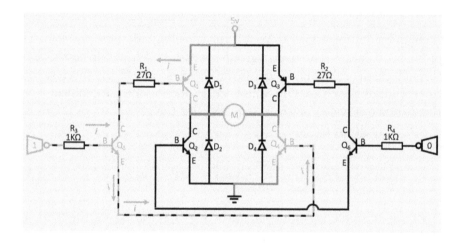

Figure A.2 Schematic design of the implemented H-bridge.

that, together with a horseshoe-type optical sensor ($H21A1$), allowed a pulse counter to be obtained per revolution.

Figure A.3 shows the designed testbed (the motor is in red, and the components of the base are in white), with the shaft secured to a 2-mm-thick disk (disk in blue) that passes through an optical horseshoe sensor (gray device) fixed to the same base, so that the RPM can be measured. The dimensions of the motor and optical sensor are found in [44] and [73], respectively.

A low-pass filter was implemented into the circuit according to the maximum input and maximum values of 5 V and 5290 RPM, respectively:

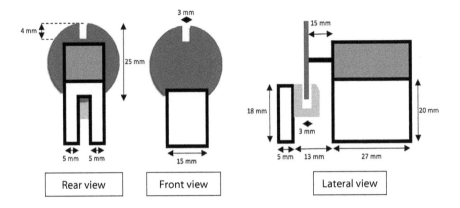

Figure A.3 Designed testbed.

$$f_{max} = \frac{5300\ RPM\ *\ 1\ Hz}{60\ RPM} = 83.33\ Hz. \tag{A.2}$$

where f_{max} is the maximum frequency expected. The circuit schematic and board designs using the components in [33,44,70–73] are presented in Figures A.4 and A.5, respectively.

In Figure A.4, the jack connections at the left are used for data acquisition and signal control. The pins at the right are used for the motor voltage and optical sensor output.

Figure A.5 shows the implemented circuit board design taken from the schematic shown in Figure A.4, which also clearly shows a space without components, which represents the same space that is used to fix the motor base as shown in Figure A.3 to the circuit board.

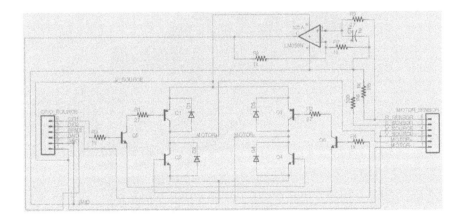

Figure A.4 Schematic of the implemented test circuit.

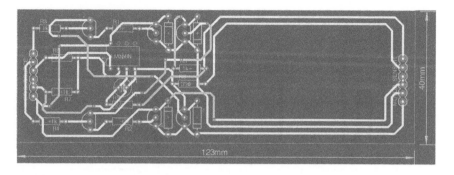

Figure A.5 Board design of the implemented test circuit.

The complete designed test environment made for experimental speed analysis and control and implemented with the *PPN13KB DC* motor [44] is shown in Figure A.6.

Figure A.6 Testing environment.

Bibliography

[1] Bilal Alatas. ACROA: Artificial chemical reaction optimization algorithm for global optimization. *Expert Systems with Applications*, 38(10):13170–13180, 2011.

[2] Michael Allaby. *Dictionary of Geology and Earth Sciences*. Oxford University Press, Oxford, 2013.

[3] Mohammed R AlRashidi and Mohamed E El-Hawary. A survey of particle swarm optimization applications in electric power systems. *IEEE transactions on evolutionary computation*, 13(4):913–918, 2008.

[4] BetaCommunityCo. LV FPGA pseudo random number generator. https://forums.ni.com/t5/Example-Programs/LV-FPGA-Pseudo-Random-Number-Generator/ta-p/3492970?profile. language=es, 2007. Accessed in 2019.

[5] Jeffery Y Beyon. *LabVIEW Programming, Data Acquisition and Analysis*. Prentice Hall PTR, Upper Saddle River, NJ, 2000.

[6] Rick Bitter, Taqi Mohiuddin, and Matt Nawrocki. *LabVIEW: Advanced Programming Techniques*. Crc Press, Boca Raton, FL, 2017.

[7] Peter Bormann. *New Manual of Seismological Observatory Practice (NMSOP-2)*, volume 2 Volumes. 01 2002.

[8] Peter Bormann, Bob Engdahl, and Kind Rainer. Seismic wave propagation and earth models. 2009.

[9] Young-kyu Choi, Jason Cong, Zhenman Fang, Yuchen Hao, Glenn Reinman, and Peng Wei. A quantitative analysis on microarchitectures of modern CPU-FPGA platforms. In *Proceedings of the 53rd Annual Design Automation Conference*, DAC '16, pages 109:1–109:6, New York, NY, USA, 2016. ACM. event-place: Austin, Texas.

[10] John CB Cooper. The poisson and exponential distributions. *Mathematical Spectrum*, 37(3):123–125, 2005.

[11] Matej Črepinšek, Shih-Hsi Liu, and Marjan Mernik. Exploration and exploitation in evolutionary algorithms: A survey. *ACM Computing Surveys (CSUR)*, 45(3):35, 2013.

[12] Sandor Csiszar. Optimization algorithms (survey and analysis). In *2007 International Symposium on Logistics and Industrial Informatics*, pages 185–188. IEEE, 2007.

[13] Gaurav Dhiman and Vijay Kumar. Spotted hyena optimizer: A novel bio-inspired based metaheuristic technique for engineering applications. *Advances in Engineering Software*, 114:48–70, December 2017.

[14] Marco Dorigo and Mauro Birattari. *Ant colony optimization*. Springer, New York, 2010.

[15] Marco Dorigo and Mauro Birattari. Ant colony optimization. In *Encyclopedia of Machine Learning*, pages 36–39. Springer, New York, 2011.

[16] Marco Dorigo, Vittorio Maniezzo, and Alberto Colorni. Positive feedback as a search strategy. 1991.

[17] Kathryn A Dowsland and Jonathan M Thompson. Simulated annealing. In: Rozenberg, Grzegorz, Back, Thomas and Kok, Joost N. eds. *Handbook of Natural Computing*, pages 1623–1655, Springer-Verlag, 2012.

[18] Haifeng Du, Xiaodong Wu, and Jian Zhuang. Small-world optimization algorithm for function optimization. In *International Conference on Natural Computation*, pages 264–273. Springer, Berlin, Heidelberg, 2006.

[19] Russ C Eberhart, James Kennedy, et al. A new optimizer using particle swarm theory. In *Proceedings of the Sixth International Symposium on Micro Machine and Human Science*, volume 1, pages 39–43. New York, NY, 1995.

[20] Emad Elbeltagi, Tarek Hegazy, and Donald Grierson. Comparison among five evolutionary-based optimization algorithms. *Advanced Engineering Informatics*, 19(1):43–53, 2005.

[21] Eid Emary, Hossam M Zawbaa, and Crina Grosan. Experienced gray wolf optimization through reinforcement learning and neural networks. *IEEE Transactions on Neural Networks and Learning Systems*, 29(3):681–694, 2017.

[22] John Fitzgerald, Peter Gorm Larsen, and Marcel Verhoef. Collaborative Design for Embedded Systems. *Springer Berlin Heidelberg, Berlin, Heidelberg. http://link.springer.com/10.1007/978-3-642-54118-6 DOI*, 10:978–3, 2014.

[23] Lawrence J Fogel, Alvin J Owens, and Michael J Walsh. Artificial intelligence through simulated evolution. 1966.

[24] Amir Hossein Gandomi, Xin-She Yang, and Amir Hossein Alavi. Cuckoo search algorithm: A metaheuristic approach to solve structural optimization problems. *Engineering with Computers*, 29(1):17–35, January 2013.

[25] Zong Woo Geem, Joong Hoon Kim, and Gobichettipalayam Vasudevan Loganathan. A new heuristic optimization algorithm: harmony search. *Simulation*, 76(2):60–68, 2001.

[26] Maria George and Peter Alfke. Linear feedback shift registers in virtex devices. page 5, 2007.

[27] Fred Glover. Tabu search, part I. *ORSA Journal on Computing*, 1(3):190–206, 1989.

[28] Fred Glover and Manuel Laguna. Tabu search. In *Handbook of Combinatorial Optimization*, pages 2093–2229. Springer, 1998.

[29] Haluk Gozde, M. Cengiz Taplamacioglu, and İlhan Kocaarslan. Comparative performance analysis of artificial bee colony algorithm in automatic generation control for interconnected reheat thermal power system. *International Journal of Electrical Power & Energy Systems*, 42(1):167–178, 2012.

[30] Abdolreza Hatamlou. Black hole: A new heuristic optimization approach for data clustering. *Information Sciences*, 222:175–184, Elsevier 2013.

[31] N. Hemsoth and M. Prickett. *FPGA Frontiers: New Applications in Reconfigurable Computing*. Next Platform Press, 2017.

[32] John H Holland. Adaptation in Natural and Artificial Systems. An Introductory Analysis with Application to Biology, Control, and Artificial Intelligence. *Ann Arbor, MI: University of Michigan Press*, pages 439–444, 1975.

[33] Texas Instruments. Lmx58-n low-power, dual-operational amplifiers, Data Sheet. Texas Instruments, 2014.

[34] Momin Jamil and Xin-She Yang. A literature survey of benchmark functions for global optimisation problems. *International Journal of Mathematical Modelling and Numerical Optimisation*, 4(2):150–194, 2013.

[35] Dervis Karaboga and Bahriye Basturk. A powerful and efficient algorithm for numerical function optimization: artificial bee colony (ABC) algorithm. *Journal of Global Optimization*, 39(3):459–471, 2007.

[36] Tuba Karaboga, Murat Canyilmaz, and Osman Ozcan. Investigation of the relationship between ionospheric foF2 and earthquakes. *Advances in Space Research*, 61(8):2022–2030, 2018.

[37] Ali Husseinzadeh Kashan. League championship algorithm (LCA): An algorithm for global optimization inspired by sport championships. *Applied Soft Computing*, 16:171–200, 2014.

[38] A Kaveh and S Talatahari. A novel heuristic optimization method: Charged system search. *Acta Mechanica*, 213(3-4):267–289, 2010.

[39] James Kennedy. Particle swarm optimization. In Sammut C., Webb G.I. (eds) *Encyclopedia of Machine Learning*, Springer, Boston, MA, pages 760–766, 2010.

[40] Scott Kirkpatrick, C Daniel Gelatt, and Mario P Vecchi. Optimization by simulated annealing. *Science*, 220(4598):671–680, 1983.

[41] Thorne Lay. A review of the rupture characteristics of the 2011 tohoku-oki mw 9.1 earthquake. *Tectonophysics*, 733, 2017.

[42] Wei Liu and Jianyu Wang. A brief survey on nature-inspired metaheuristics for feature selection in classification in this decade. In *2019 IEEE 16th International Conference on Networking, Sensing and Control (ICNSC)*, pages 424–429. IEEE, 2019.

[43] Manuel García López, Pedro Ponce, Luis Arturo Soriano, Arturo Molina, and Jaime José Rodríguez Rivas. A novel fuzzy-pso controller for increasing the lifetime in power electronics stage for brushless dc drives. *IEEE Access*, 7:47841–47855, 2019.

[44] Minebea Co. Ltd. Dc brush motor ppn13, 2010.

[45] Efrain Mendez, Alexandro Ortiz, Pedro Ponce, Juan Acosta, and Arturo Molina. Mobile phone usage detection by ann trained with a metaheuristic algorithm. *Sensors*, 19(14):3110, 2019.

[46] Seyedali Mirjalili, Seyed Mohammad Mirjalili, and Andrew Lewis. Grey wolf optimizer. *Advances in Engineering Software*, 69:46–61, 2014.

[47] Cleve B. Moler. Numerical Computing with MATLAB: Revised Reprint, SIAM, 87, 2008.

[48] Marcin Molga and Czesław Smutnicki. Test functions for optimization needs. *Test Functions for Optimization Needs*, 101, 2005.

[49] James A Momoh, Rambabu Adapa, and ME El-Hawary. A review of selected optimal power flow literature to 1993. i. nonlinear and quadratic programming approaches. *IEEE Transactions on Power Systems*, 14(1):96–104, 1999.

[50] Antonio Mucherino and Onur Seref. Monkey search: A novel metaheuristic search for global optimization. In *AIP Conference Proceedings*, volume 953, pages 162–173. AIP, 2007.

[51] John A Nelder and Robert Mead. The downhill simplex algorithm. *Computer Journal*, 7(S 308), 1965.

[52] A. Foroughi Nematollahi, A. Rahiminejad, and B. Vahidi. A novel physical based meta-heuristic optimization method known as lightning attachment procedure optimization. *Applied Soft Computing*, 59:596–621, 2017.

[53] NI. Daqmx express vi tutorial. http://www.ni.com/tutorial/2744/en/, 2019. Accessed in 2019.

[54] NI. A global leader in automated test and automated measurement systems. https://www.ni.com/en-us.html, 2019. Accessed in 2019.

[55] NI. Ni community. https://forums.ni.com/?profile.language=en, 2019. Accessed in 2019.

[56] NI. Shop - national instruments. https://www.ni.com/en-us/shop.html, 2019. Accessed in 2019.

[57] Xiaohong Nian, Fei Peng, and Hang Zhang. Regenerative braking system of electric vehicle driven by brushless dc motor. *IEEE Transactions on Industrial Electronics*, 61(10):5798–5808, 2014.

[58] Jorge Nocedal and Stephen Wright. *Numerical Optimization*. Springer Science & Business Media, 2006.

[59] Erik L Olson and Richard M Allen. The deterministic nature of earthquake rupture. *Nature*, 438(7065):212, 2005.

[60] Ponce Pedro and Molina Arturo. Earthquake optimization algorithm, internal report binational laboratory project, 2017.

[61] Pragasen Pillay and Ramu Krishnan. Modeling, simulation, and analysis of permanent-magnet motor drives. ii. the brushless dc motor drive. *IEEE Transactions on Industry Applications*, 25(2):274–279, 1989.

[62] Pragasen Pillay and Ramu Krishnan. Application characteristics of permanent magnet synchronous and brushless dc motors for servo drives. *IEEE Transactions on Industry Applications*, 27(5):986–996, 1991.

[63] Pedro Ponce and Arturo Molina. Internal report project - 266636 grant. Binational Laboratory Mexico-Conacyt-Tecnologico de Monterrey, 2018.

[64] Pedro Ponce-Cruz, Arturo Molina, and Brian MacCleery. *Fuzzy Logic Type 1 and Type 2 Based on LabVIEW^{TM} FPGA*. Springer, 2016.

[65] Rajavel Rajakumar, Ponnurangam Dhavachelvan, and Thirumal Vengattaraman. A survey on nature inspired meta-heuristic algorithms with its domain specifications. In *2016 International Conference on Communication and Electronics Systems (ICCES)*, pages 1–6. IEEE, 2016.

[66] R.V. Rao, V.J. Savsani, and D.P. Vakharia. Teaching–learning-based optimization: A novel method for constrained

mechanicaldesign optimization problems. *Computer-Aided Design*, 43(3):303–315, 2011.

[67] Esmat Rashedi, Hossein Nezamabadi-pour, and Saeid Saryazdi. Gsa: A gravitational search algorithm. *Information Sciences*, 179(13):2232–2248, 2009. Special Section on High Order Fuzzy Sets.

[68] Ingo Rechenberg. Cybernetic solution path of an experimental problem. 1965.

[69] Michael Rüßmann, Markus Lorenz, Philipp Gerbert, Manuela Waldner, Jan Justus, Pascal Engel, and Michael Harnisch. Industry 4.0: The future of productivity and growth in manufacturing industries. *Boston Consulting Group*, 9(1):54–89, 2015.

[70] F. Semicond. 2n3906 data sheet, 2000.

[71] F. Semiconducto. 1n4001-4007 general purpose rectifiers, 2009.

[72] F. Semiconductor. 2n3904/mmbt3904/pzt3904 npn general purpose amplifier, 2014.

[73] Fairchild Semiconductors. H22a1 slotted optical switch, 2010.

[74] Hamed Shah-Hosseini. Principal components analysis by the galaxy-based search algorithm: A novel metaheuristic for continuous optimisation. *International Journal of Computational Science and Engineering*, 6(1-2):132–140, 2011.

[75] Saša Singer and John Nelder. Nelder–Mead algorithm. *Scholarpedia*, 4(7):2928, 2009.

[76] Rainer Storn and Kenneth Price. Differential evolution–a simple and efficient heuristic for global optimization over continuous spaces. *Journal of Global Optimization*, 11(4):341–359, 1997.

[77] S. Talatahari, B. Farahmand Azar, R. Sheikholeslami, and A.H. Gandomi. Imperialist competitive algorithm combined with chaos for global optimization. *Communications in Nonlinear Science and Numerical Simulation*, 17(3):1312–1319, 2012.

[78] Mohammad-H Tayarani-N and MR Akbarzadeh-T. Magnetic optimization algorithms a new synthesis. In *IEEE Congress on Evolutionary Computation, 2008 (IEEE World Congress on Computational Intelligence).*, pages 2659–2664. IEEE, 2008.

[79] Jeffrey Travis and Jim Kring. *LabVIEW for Everyone: Graphical Programming Made Easy and Fun*. Prentice-Hall, Upper Saddle River, NJ, 2007.

[80] UNAM. Seismic data provided by the strong ground motion database system. http://aplicaciones.iingen.unam.mx/AcelerogramasRSM/i, 2014. Accessed in feb-2018.

[81] Peter JM Van Laarhoven and Emile HL Aarts. Simulated nnealing. In *Simulated Annealing: Theory and Applications*, pages 7–15. Springer, 1987.

[82] John GW West. Dc, induction, reluctance and pm motors for electric vehicles. *Power Engineering Journal*, 8(2):77–88, 1994.

[83] Chenguang Yang, Xuyan Tu, and Jie Chen. Algorithm of marriage in honey bees optimization based on the wolf pack search. In *Intelligent Pervasive Computing, 2007. IPC. The 2007 International Conference on*, pages 462–467. IEEE, 2007.

[84] Xin-She Yang, Suash Deb, Simon Fong, Xingshi He, and Yu-Xin Zhao. From Swarm intelligence to metaheuristics: Nature-inspired optimization algorithms. *Computer*, 49(9):52–59, September 2016.

[85] Xin-She Yang. Firefly algorithms for multimodal optimization. In *International Symposium on Stochastic Algorithms*, pages 169–178. Springer, Berlin, Heidelberg, 2009.

[86] Xin-She Yang. *Nature-Inspired Metaheuristic Algorithms*. Luniver press, 2010.

[87] Xin-She Yang. A new metaheuristic bat-inspired algorithm. In *Nature Inspired Cooperative Strategies for Optimization (NICSO 2010)*, pages 65–74. Springer, Berlin, Heidelberg, 2010.

[88] Xin-She Yang and Amir Hossein Gandomi. Bat algorithm: A novel approach for global engineering optimization. *Engineering Computations*, 29(5):464–483, 2012.

Index

A

Ant Colony Optimization
(ACO), 4, 5, 71, 77, 88,
92, 95, 100, 101, 109,
111, 124, 125, 126, 128,
129
attraction level, 71
pheromone trail, 71
route, 71, 86, 111, 115,
126, 143
Arrays, 6, 21, 22, 53, 96, 111,
118, 132

B

Bat Optimization Algorithm
(BA), 5, 73, 74, 77, 88,
89, 90, 94, 100, 128,
133, 142
bats, 73, 74, 89, 94
constants, 21, 74, 89
echolocation, 73, 74
emission rate, 74, 89, 94
frequency, 74, 89, 94, 118,
121, 122, 145
loudness, 74, 89, 94
position, 7, 46, 73, 74, 75,
76, 84, 93, 94, 95, 96,
113, 114, 125
velocity, 46, 73, 74, 79, 80,
84, 89, 94, 113, 116
wavelength, 74
Bio-inspired, 74
Blackbox optimization, 61

Benchmark functions, 99, 100,
101, 102, 133, 134, 135,
136, 137, 138, 141
Ackley function, 134
Adjiman function, 136, 141
Bukin n.6 function, 135,
137, 141
drop-wave function,
137, 141
egg-holder function,
136, 141
Keane function, 100, 102,
136, 138, 140, 141
Levy function, 137, 141
Rastrigin function, 136, 141
Schaffer n.2 function,
136, 141
sphere function, 99, 100,
136, 141, 142
Styblinski–Tang function,
99, 100, 101, 102,
137, 141

C

Classification, 4, 5, 78
Constants, 21, 74, 89
Constraint function, 16
Controllers, 7, 49
proportional integral
derivative (PID), 7, 82,
127, 128, 129, 130,
131, 132
Conventional optimization,
9, 16

Conventional optimization (*cont.*)
  gradient descent method, 16
  gradient vector, 14, 15
  Hessian matrix, 14, 15
  local minimize, 14
Cost function, 8, 9, 10, 95, 143

D
Dataset, 4, 5
DC/DC converter, 112
  boost converter, 112,
    116, 118
  duty cycle, 112, 113, 114,
    115, 118, 128, 129, 130
  power converter, 111, 112
DC motor speed controller
    tuning optimization,
    127, 130, 131, 132
  performance indexes, 128
  remote application, 132
  testbed, 122
Dynamic optimization, 112,
    115, 119, 120

E
Earthquake algorithm (EA), 4,
    78, 83, 88, 114
  earth material, 79, 80, 81,
    82, 83
  epicenters, 80, 84, 86, 89,
    91, 95, 115, 120
  exponential distribution,
    84, 86
  Lamé parameters, 83
  P-wave velocity, xiii, 79, 80,
    81, 83
  poisson ratio, 79, 82, 83
  position, 7, 46, 67, 73, 74,
    75, 76, 84, 86, 93, 94,
    95, 96, 113, 114,
    115, 125

S-range (Sr), 84, 115
S-wave, 80, 81, 82, 83
  velocity, 46, 73, 74, 79, 80,
    84, 89, 94, 113
Evolutionary algorithms, 61
Exhaustive search, 61

F
Field-programmable gate array
    (FPGA), 6, 8, 16, 17,
    130, 132
  benchmark functions, 99,
    100, 101, 102, 133, 134,
    135, 136, 137, 138, 141
  implementation, 8, 34, 42,
    43, 108, 140, 142, 144,
    145, 146
  metaheuristic algorithms,
    4, 5, 6, 77, 118, 120,
    129, 132, 133
  national instruments
    cRIO-9030, 133
  pseudo random number
    generation, 141
Fuzzy logic, 6, 7, 8, 9
  fuzzy logic controller
    optimized by PSO, 9

G
Genetic algorithms (GA), 4, 5,
    65, 66, 77, 88, 89, 90,
    93, 99
  characteristics, 6, 65, 77,
    133
  crossing, 65, 66, 93
  generation, 6, 65, 66, 86,
    89, 111, 121, 140, 141
  mutation, 65, 66, 89, 93
  reproduction, 65
Geo-inspired, 78, 82, 114

Geological optimization, 77, 79,
81, 83, 85
Graphs, 6, 36, 39, 89, 92
LabVIEW, 12, 14, 20, 21,
47, 48, 49, 50, 51, 53,
54, 56, 87, 96, 98, 99,
100, 101, 102, 125, 127,
130, 131, 132
MATLAB, 12, 13, 17, 18,
19, 20, 21, 22, 24, 25,
26, 28, 29, 33, 34, 35,
36, 45, 56, 87, 96, 124
Gray Wolf Optimization
(GWO), 74, 75, 76, 88,
89, 90, 95, 101, 126,
128, 129, 133, 142
Alpha Wolf, 75
attacking the prey, 75
beta, delta and omega
wolves, 75
coefficients, 63, 75, 92, 119
encircling and harassing
the prey, 75, 76
hunting process, 74, 75, 76
searching for pray
(tracking), 75, 76
Group Counseling Optimization
(GCO), 5

H

Heuristic optimization, 3, 4
methodology, 3, 4, 5

I

Industry 4.0, 121
LabVIEW solution, 125
MATLAB solution, 124
three-phase inverter case
study, 121

L

LabVIEW, 12, 14, 20, 21, 47,
48, 49, 50, 51, 53, 54,
56, 87, 96, 98, 99, 100,
101, 102, 125, 127, 130,
131, 132
block diagram, 48, 49, 128,
141
front panel, 48, 49, 53, 56,
96, 98, 99, 100, 101,
102, 130
fundamentals, 49
LabVIEW toolkit, 87, 89, 91,
93, 95, 97, 99, 101
ACO, 5, 71, 77, 88, 92, 95,
100, 101, 109, 111, 124,
125, 126, 128, 129
app, 87, 88, 92, 96, 98, 103,
109, 111, 125, 130, 131
BA, 5, 77, 88, 89, 90, 94,
100, 128, 133, 142
EA, 7, 13, 14, 15, 78, 82,
83, 84, 85, 86, 88, 89,
90, 95, 101, 114, 115,
116, 117, 120, 128, 129,
133, 142
front panels, 96
GA, 5, 65, 66, 77, 88, 89,
90, 93, 99
GWO, 74, 75, 76, 88, 89,
90, 95, 101, 128, 129,
133, 142
NM, 88, 89, 90, 92, 102,
133, 142
PSO, 4, 5, 8, 9, 72, 73, 77,
88, 89, 90, 94, 99, 113,
114, 115, 116, 120, 128,
129, 133, 142

LabVIEW toolkit (*cont.*)
    simulated annealing, 4, 5, 66, 67, 77, 88, 89, 90, 93
    Tabu search, 4, 5, 67, 68, 69, 77, 88, 92, 93, 95
Learning algorithm, 5
Linear square regression, 107
Low-pass filter, 144

M

Mass-Spring-Damper system, 43, 46
    Simulink model, 43, 45, 116
MATLAB, 12, 13, 17, 18, 19, 20, 21, 22, 24, 25, 26, 28, 29, 33, 34, 35, 36, 45, 56, 87, 96, 124
    code, 36, 38, 39, 54, 55, 56, 61, 75, 82, 83, 144
    command window, 18, 20, 25, 26, 28, 43
    editor, 28
    fundamentals, 17, 43
    symbolic code, 10, 12
    workspace, 18, 19, 20, 25, 26, 44, 45
MATLAB app
    ACO, 5, 71, 77, 88, 92, 95, 100, 101, 109, 111, 124, 125, 126, 128, 129
    app, 87, 88, 92, 96, 98, 103, 109, 111, 125, 130, 131
    BA, 5, 77, 88, 89, 90, 94, 100, 128, 133, 142
    EA, 7, 13, 14, 15, 78, 82, 83, 84, 85, 86, 88, 89, 90, 95, 101, 114, 115, 116, 117, 120, 128, 129, 133, 142

GA, 5, 65, 66, 77, 88, 89, 90, 93, 99
GWO, 74, 75, 76, 88, 89, 90, 95, 101, 128, 129, 133, 142
interfaces, 88, 89, 91
NM, 88, 89, 90, 92, 102, 133, 142
PSO, 4, 5, 8, 9, 72, 73, 77, 88, 89, 90, 94, 99, 113, 114, 115, 116, 120, 128, 129, 133, 142
simulated annealing, 4, 5, 66, 67, 77, 88, 89, 90, 93
Tabu search, 4, 5, 67, 68, 69, 77, 88, 92, 93, 95
Maximum power point tracking, 112
    case study, 6, 111, 112, 114, 119, 121, 122, 123, 128
    EA MPPT, 116, 117
    MPPT, 96, 97, 98, 111, 112, 113, 114, 115, 116, 117, 118, 120
    Perturb and Observe (P&O), 96, 112
    PSO MPPT, 115
    random generation, 86
Memetic algorithms, 71, 73, 75
Metaheuristic optimization, 4, 5, 6, 16, 57, 59, 61, 63, 78, 82, 114
    global best solution, 4, 84
    local best solution, 86
    local search, 4, 5, 67, 74
    randomization, 4
Motors, 6, 7, 121, 122
    alternating current (AC) motors, 6

brushless direct current
motors (BLDCM), 6,
7, 121, 122
direct current (DC) motors,
6
Multi-objective
optimization, 15

N

Nelder–Mead algorithm (NM),
61, 62, 63, 64, 88, 133
contraction, 17, 62, 63,
79, 92
expansion, 62, 63, 92
shrunk, 62, 63, 92

O

Objective function, 3, 8, 9, 10,
14, 15, 16, 65, 73, 89,
96, 128, 130, 131, 140,
142
Optical sensor, 129, 144, 145
Optimization, 3, 4, 5, 6, 7, 8, 9,
15, 16, 17, 57, 59, 61,
62, 65, 66, 71, 72, 73,
74, 77, 78, 82, 83, 87,
88, 89, 90, 92, 93, 94,
95, 96, 99, 100, 101,
102, 109, 111, 112, 113,
114, 115, 119, 120, 122,
123, 124, 125, 126, 127,
128, 129, 130, 131, 132,
133, 134, 136, 137, 140,
141, 142, 143
global minimum, 10, 14,
103, 136
global maximum, 10, 14,
103, 105
local minimum, 10, 14,
86, 103

local maximum, 10, 14,
103, 105
maximization, 10, 97
minimization, 8, 10, 108
newton-based
optimization, 3
Optimization algorithms, 3, 8,
16, 61, 63, 78, 90, 96,
114, 127, 128, 139, 132,
133, 134, 140, 142
exploitation, 3, 76
exploration, 3, 76, 79
Optimization functions, 3
Optimization problem, 3, 4, 8,
9, 15, 143

P

Particle swarm optimization
(PSO), 4, 5, 8, 9, 72,
73, 77, 88, 89, 90, 94,
99, 113, 114, 115, 116,
120, 128, 129, 133, 142
coefficients, 114, 120
inertia, 73, 77, 89, 94, 120
particles, 4, 5, 72, 73, 77,
81, 82, 84, 88, 89, 94,
113, 114, 120
position, 7, 46, 67, 73, 74,
75, 76, 84, 93, 94, 95,
96, 113, 114, 115, 125
swarm, 4, 5, 72, 77, 78, 88,
113
swarm intelligence, 4
velocity, 46, 73, 74, 79, 80,
84, 89, 94, 113, 116
weight, 6, 73, 89, 94, 120
Photovoltaic (PV) systems,
96, 111
irradiation, 111, 114, 120
load, 25, 26, 111, 112, 118

Photovoltaic (PV) systems
(*cont.*)
    power converter, 111, 112
    PV array, 111, 112, 113,
       118, 119
Power electronics, 111, 121
    field-effect transistor
       (FET), 7
    insulated-gate bipolar
       transistor (IGBT), 7
Printed circuit board (PCB),
       122, 123, 125, 127
    DC motor testbed, 122
    three-phase inverter, 121

R

Random optimization, 61, 62

S

Seeker optimization algorithm
       (SOA), 5
Simulated annealing, 4, 5, 66,
       67, 77, 88, 89, 90, 93
    local optima, 4
    probability factor, 4
    process of metals, 4, 66
Simulink, 43, 44, 45, 46, 47, 96,
       97, 116, 119
    blocks, 32, 43, 44, 45, 46,
       48, 49, 50, 51, 56, 112,
       118, 128, 141, 142
    fundamentals, 3, 8, 17, 43,
       49
    signals, 7, 8, 43, 44, 45,
       129, 130, 133, 145
Social-based algorithm
       (SBA), 5
Stochastic decision, 61

T

Tabu search, 4, 5, 67, 68, 69,
       77, 88, 92, 93, 95
    aspiration, 68
    tabu, 4, 5, 67, 68, 69, 77,
       88, 92, 93, 95
Traveling salesman
       problem, 93, 109,
       110, 124, 143
    3D traveling salesman
       problem, 109
    traveling salesman
       person, 143
Testbed design, 122, 144, 145
Turing, A., 4
    heuristic search, 4

V

Variables definition, 19
    code, 36, 38, 39, 54, 55, 56,
       61, 75, 82, 83, 144
    LabVIEW, 12, 14, 20, 21,
       47, 48, 49, 50, 51, 53,
       54, 56, 87, 96, 98, 99,
       100, 101, 102, 125, 127,
       130, 131, 132
    MATLAB, 12, 13, 17, 18,
       19, 20, 21, 22, 24, 25,
       26, 28, 29, 33, 34, 35,
       36, 45, 56, 87, 96, 124
Vectors, 19, 21, 22, 75

W

Welded cantilever
       minimization, 108